Decolonizing Our Names in the 21st Century

This book combines different decolonial approaches from around the world to offer a roadmap for updating names and naming practices, restoring and protecting precolonial ones, and reimagining or recontextualizing the relationship between place, identity, and names.

In a postcolonial context, naming often serves as a bitter reminder of past harms through commemorative naming practices, whether through a system of baptismal names or a former colony's approach to dealing with the names that the colonizer left behind. This volume assembles authors who hail from formerly colonized regions of the Americas, Europe, Africa, and Asia to engage with this problem of decolonizing names in the twenty-first century from a global perspective. The book also points to what strategies have had more success than others while envisioning the tools needed for progress in the future.

Offering a useful framework with approaches that can easily be used across other geographical contexts, this volume is suitable for scholars and students interested in decolonization, identity, and naming practices.

Lauren Beck is a professor of visual and material culture studies at Mount Allison University, Canada, and specializes in place name science and identity. Her publications include *Canada's Place Names and How to Change Them* and *Firsting in the Early Modern Transatlantic World.*

Grace A. Gomashie is a postdoctoral researcher at the Centre for Early Modern Visual Culture, Mount Allison University, Canada, where she researches topics on equity, diversity, inclusion, and decolonization. She has published in onomastics, Indigenous language maintenance, Spanish varieties, community studies, and translation studies.

Routledge Studies in Modern History

Building Europe in New York
From the Munich Conference to the European Coal and Steel Community (1938–1952)
Enrico Ciappi

The Psychology of the Armenian Genocide
George Green

"Dangerous Vagabonds"
French Senegal and Resistance to Slave Emancipation
Robin A. Hardy

(Post)Colonial Ports
Place and (Non)Place in the Ecotone
Edited by Jill Didur and Nalini Mohabir

Gender, Women and the Indian Emergency, 1975–1977
Gemma Scott

Memory and the Holocaust
Descendants of Survivors and Family History
Edited by Judith Tydor Baumel-Schwartz and David Clark

Technical and Scientific Training in the Construction of Empires
On the Quest of Learning Places
Edited by Alice Santiago Faria, Renata Malcher de Araujo and Margarida Tavares da Conceição

Decolonizing Our Names in the 21st Century
Place, Identity, and Agency
Edited by Lauren Beck and Grace A. Gomashie

For more information about this series, please visit: https://www.routledge.com/Routledge-Studies-in-Modern-History/book-series/MODHIST

Decolonizing Our Names in the 21st Century

Place, Identity, and Agency

Edited by Lauren Beck and Grace A. Gomashie

Routledge
Taylor & Francis Group
NEW YORK AND LONDON

First published 2026
by Routledge
605 Third Avenue, New York, NY 10158

and by Routledge
4 Park Square, Milton Park, Abingdon, Oxon, OX14 4RN

Routledge is an imprint of the Taylor & Francis Group, an informa business

ISBN: 978-1-032-54021-4 (hbk)
ISBN: 978-1-032-54022-1 (pbk)
ISBN: 978-1-003-41477-3 (ebk)

DOI: 10.4324/9781003414773

Typeset in Times New Roman
by Apex CoVantage, LLC

Contents

Figures

Tables

Acknowledgements

We would like to recognize the financial support of the Social Sciences and Humanities Research Council of Canada, the Canada Research Chair Secretariat, and Mount Allison University. We are also grateful for the editorial guidance of Max Novick and the Routledge team.

Contributors

Osei Yaw Akoto is a lecturer in the Department of English at Kwame Nkrumah University of Science and Technology, Ghana. His research interests include linguistic landscape, onomastics, Corpus Linguistics and academic discourse and he is the founding president of Ghana Name Society.

Lauren Beck is a professor of visual and material culture studies and the Canada research chair in intercultural encounter at Mount Allison University, Canada. She researches the early modern visual culture of the Atlantic world and has published several monographs, including *Canada's Place Names and How to Change Them* (2022), as well as edited collections, among them *Firsting in the Early Modern Transatlantic World* (2019). She is the series editor of the six-volume *Cultural History of Exploration* (2024) and has previously held editorial positions at *Terrae Incognitae*, among other venues.

Grace A. Gomashie is a postdoctoral researcher at the Centre for Early Modern Visual Culture, Mount Allison University, Canada, where she researches topics on equity, diversity, inclusion, and decolonization. She holds a PhD in Hispanic studies from University of Western Ontario, Canada, where her doctoral research focused on language vitality in Indigenous communities. She has published peer-reviewed articles on onomastics, Indigenous language maintenance, Spanish varieties, community studies, and translation studies. Her research interest in onomastics began with her master's thesis on the sociolinguistic study of Fang names in Equatorial Guinea (2014). Since 2019, she has been an executive member of the Canadian Society for the Study of Names. She is currently a member of the editorial board of *Onomastica Canadiana*.

Federica Guccini obtained her PhD in Anthropology and a collaborative specialization in Migration and Ethnic Relations from the University of Western Ontario (Western University), Canada in 2022. Her doctoral research on Hakka Chinese Mauritian migration, language and identity practices was supported by a Vanier Canada Graduate Scholarship. She has studied names and naming practices from an anthropological perspective since 2015.

Jerrilyn McGregory is a professor of folklore at Florida State University. She is the author of *Wiregrass Country, Downhome Gospel: African American Spiritual Activism in Wiregrass Country* and *One Grand Noise: Boxing Day in the Anglicized Caribbean World*. A recipient of the Popular Culture Association's Harry Shaw Award for Outstanding Contributions to the Study of African American Popular Culture, her recent paper presentations include "Stop Being So Damn Creative" and " 'Ghetto Names': Popular Culture's Devaluation of African American Personal Names." Most recently, in 2020, the paper "Pronounce It like It Sounds: The Devaluation of African American Names" was delivered at the American Names Society conference in New Orleans.

Anna Mongibello is a senior tenure-track researcher in English language and translation at the University of Naples "L'Orientale," Italy. Her research interests include the intersections of language, ideology, and identity in the Canadian context, with a focus on the practices of representation/self-representation of Indigenous peoples in Canada. She has published numerous articles in national and international journals. She is the author of two books: *Indigenous Peoples in Canadian TV News: A Corpus-Based Analysis of Mainstream and Indigenous News Discourses* (2018) and *Geografie alternative: Scrittrici indigene contemporanee del Canada anglofono* (2013).

Maxwell Mpotsiah teaches English Language and Literature-in-English in Ahanta-West Municipal, Ghana. He holds master of Arts in English Language and Bachelor of Education in English Language from the University of Education, Winneba. His areas of interest include English grammar, stylistics, pragmatics, World Englishes, discourse analysis, sociolinguistics, and literature.

Joy Aruoture Omoru is a lecturer in the Department of Languages and Linguistics, Anchor University Lagos, Nigeria. She holds master's and PhD degrees in English with a research focus on oral literatures and cultural science. She is a member of the Nigerian Oral Literature Association (NOLA) and other professional bodies. She has served as a deputy editor of *Anchor Magazine* and been on the editorial board of the *Nigerian Journal of Oral Literature*. She has participated in several local and international conferences and has also published several articles in local and international journals and book chapters.

Israel Abayomi Saibu is a PhD candidate in African History and lecturer in the Department of History and International Studies, Anchor University, Lagos, Nigeria. He has published articles in both local and international journals on African history, migration dynamics, digital humanities, cultural heritage, environmental history, and international economy.

Peter K. W. Tan is a professor of English language and literature at the National University of Singapore. His research interests include onomastics in the context of Singapore and Malaysia, where he examines changing personal names and investigates how street names, building names, and other names in the linguistic landscape can represent attempts at language planning or engineering. He also has an interest in the development of non-Anglo Englishes in South-east Asia (and World Englishes in general), including their literary and computer-mediated forms, and how this development has influenced naming conventions. His publications have appeared in various books and journals.

Phùng Thị Thanh Lâm is a senior researcher of Vietnamese studies at VNU-IVIDES. She obtained a doctorate in Vietnamese studies at the VNU—Institute of Vietnamese Studies and Development Sciences in 2017. Her dissertation focuses on street names in Hanoi city, which was published in 2020 as *Street Names in Hanoi, from the Late Nineteenth Century to Present*. She researches place names and Vietnam urbanization. At present, she is conducting a project that focuses on street names and smart cities in Vietnam.

Introduction

Decolonizing Our Names in the 21st Century

Lauren Beck and Grace A. Gomashie

The last three decades have witnessed broad efforts to address the coloniality of the names that designate our communities and the people who live in or come from them. In 2007, the United Nations Group of Experts on Geographical Names (UNGEGN) Working Group on the Promotion of Indigenous and Minority Group Place Names reported 47 such projects and initiatives in 11 countries to remove offensive names, promote language retention, celebrate cultural heritage and revitalization, and support reconciliation initiatives. Building on this foundation, decolonization efforts have sought to restore precolonial Indigenous names or rename western names as seen in many countries in postcolonial Africa, such as Zimbabwe and South Africa.[1] New Zealand is now known by Aotearoa, the Maori name with which it has become synonymous. Calls to consult and give greater voice to marginalized groups who have experienced or continue to experience colonization, whether in the South Pacific, North America, Latin America, Asia, or Africa, shine light on the need to address harmful naming practices that have impacted and shaped our identities at a systemic level. In a postcolonial context—where the colonizer has left and a nation state now exists—many countries have struggled to deal with existing colonial names and naming practices, particularly when they become associated with the western condition and its intersection with economic prosperity.

This book brings a decolonial framework into dialogue with a range of disciplinary perspectives to explore how this framework can result in transformative and significant impacts on our names. Interdisciplinary fields such as place name science, have not realized until more recently the vital influence that decolonial objectives will have on toponymy for changing the character of the landscape.[2] Historians, geographers, linguists, among other disciplinary scholars, have, for the most part, not engaged with decolonial theory at the crossroads of place identity. By enjoining a decolonial framework with these disciplinary inquiries into the postcolonial toponymic status quo, we scaffold our disciplinary expertise to work toward a common objective while bringing decolonial thought to place name science. Place name science is fundamentally interdisciplinary, being a sub-discipline of onomastics researched by scholars in both the Humanities

DOI: 10.4324/9781003414773-1

(i.e., cartography, geography, history, languages, linguistics, literatures) and Sciences (i.e., environmental science, psychology).

Western influences on global naming practices remain one of the most significant challenges to decolonizing our names in former European colonies, from Canada to Nigeria and South Africa to Singapore. The paradigm of affluence and modernity, cultural and religious practices associated with the West, as well as European languages that are now spoken across the world due primarily to colonization, underline how pernicious western names can be when they supplant non-western or Indigenous ones. Systems and cultural practices governing names that range from toponyms and personal names to corporate names and virtual spaces can be inflexible, encoded in laws, charters of freedoms, and constitutions, and subject to the pressures of western forms of governance and globalization. Decolonizing names in the twenty-first century must therefore extend beyond names and their western or European origins through colonization, which implies the assertion of precolonial names or names used by people who have experienced or continue to experience colonization. In recognizing that globalization makes non-western borders porous to western naming practices—for instance, through the use of English in non-colonial contexts to propagate a place's business and tourism sectors—decolonizing names also involves a deliberate intervention into naming that reveals western values for and influences on naming, even when no colonial relationship may exist.

As the chapters contained in this book demonstrate, in the twenty-first century names are increasingly becoming an instrument of resistance to the settler-colonial normativity implied by official maps and toponyms, street signs, institutional names, and even individual and collective names given to people. In this way, some Canadians prefer to state that they come from Turtle Island (rather than Canada) or from Sikniktuk (rather than southeastern New Brunswick). The existence of more than one name for a place, entity, or concept defies the western logic undergirded by official naming boards, such as the Geographic Names Board of Canada (GNBC) or the UNGEGN, as traditionally in the West, people and places have singular and official names. The increasing flexibility with which we as a human collective approach naming in this century points to the reality that our naming vocabularies are expanding to accommodate different sources of authority, as the contentious issue over the name of the Gulf of Mexico has made plain in 2025. Furthermore, tools such as the United Nations Declaration on the Rights of Indigenous People (UNDRIP)—which many countries have adopted or are considering embracing—are transforming into calls to action so that marginalized groups choose and adopt their own names, and society more broadly subscribes to decolonized names and naming practices.

What makes this book different than others is its geographical focus on former European colonies in the Americas, Africa, and Asia, and the fact that all its authors and editors in some way contend with subaltern identities within their geographical milieux of origin and study. They bring their lived experiences to their scholarly work, whether as citizens of postcolonial countries such as Singapore and Ghana where they have experienced the evolution of naming practices or as people who have themselves been marginalized through the status quo of naming.

The Book's Organization

This collection of chapters offers both case studies that demonstrate how names are (or are not) decolonized and theorizations about decoloniality at its intersection with names and identity. The book brings together interdisciplinary scholars working in Indigenous Studies, Critical Race Studies, Women's and Gender Studies, Postcolonialism, and Onomastics, among other fields interested in decolonizing names. The book offers tools to marginalized groups around the world so that they can pursue the decolonization of their names while challenging the so-called authorities who claim to govern naming conventions and practices. Chapters and their authors hail from Africa, the Americas, Asia, and Europe, and each chapter deals with the implications that naming practices have had on a region or nation's identity. Each chapter engages in some way with one or more critical themes.

The first theme coalesces around the acts of reclaiming and reasserting precolonial names, alongside the agency of decolonizing names and the methods through which precolonial names have and can be reasserted. The first chapter in this book considers the Wet'suwet'en nation's attempt in Canada to use social media networks in order to exert authority over their ancestral lands, which are being threatened by efforts to build a pipeline through its territory. Using geotags and check-in features on social media, activists are asserting toponymy that has not been approved by the GNBC, thus offering a means of subverting settler-colonial authority over their traditional lands. The following essay looks at postcolonial Hanoi, Vietnam. After the French left in the mid-twentieth century, street names reverted to precolonial ones, whereas town and neighborhood names did not. This essay examines the persistence of Latin orthography and phonetics in Hanoi place names and how native Vietnamese populations have navigated the de-Frenchification of their names.

The second theme explored by several authors is how names are protected from colonialism using mechanisms through which Indigenous or precolonial names are and can be shielded from further colonization. Chapter 3 discusses the foreignization of Ghanaian surnames, which have served as one of the traditional markers of a person's national and ethnic identity. Ghanaian surnaming practices today show foreign influences in the adoption of western names and the westernization of Indigenous names. Analyzing a corpus of 100,000 surnames of Ghanaian university students, the chapter highlights the ongoing trend of foreignization and contributes to public discourse on preserving and using Indigenous names and decolonizing African names. The fourth chapter explores three means of protecting names from further harm through policy, education, and storytelling. The author foregrounds the Canadian toponymic landscape using concrete measures taken to decolonize it. In doing so, the essay offers tools for making decolonial change possible. Chapter 5 examines the decolonizing naming practices of the Urhobo and the Yoruba ethnic groups in Nigeria as a mechanism of identity retention and cultural resistance. For these two ethnic groups, names encode and express their histories, ideologies, worldviews, values, experiences, and narratives. Names are particularly important in these predominantly oral societies as they provide insight into the events, actions, experiences, and circumstances surrounding the birth of a child, its family, and the community. This study demonstrates the role that Indigenous names play as symbols of cultural renaissance and collective history in an African context.

The third and final theme explores how names can be reimagined or recontextualized in meaningful ways while questioning the relationship between place, identity, and names. Chapter 6 investigates approaches to decolonizing language and detangling naming practices from raciolinguistic ideologies by considering what makes a name "Chinese." It examines societal attitudes toward the adoption (or not) of non-Chinese names by Chinese migrants to reveal ideologies that (de)legitimize naming practices. The author emphasizes the creativity, agency, and decoloniality of translanguaging naming practices. The next essay explores how African American naming patterns three decades after the Civil Rights era show a turn toward decolonization and the unapologetic emancipation of naming from the white gaze. African American given names were maligned and regarded as "ghetto" and "noisy" because they did not conform to hegemonic and systemically white naming patterns–a process of resistance called unnaming. The author presents a diachronic view of African American naming practices as being representative of a decolonizing turn. The final chapter in this volume focuses on the absence of wholesale street renaming trends in post-independence Singapore. Not only does Singapore retain its colonial street names but there is also a noted increase in English-style naming trends for personal and building names. As a result, the author argues for a decolonial framework that considers other perspectives such as World Englishes, linguistic commodification, and the postmodernist appropriation of language and identity.

With this book, we make a case for rethinking the naming status quo and reflecting upon who it excludes. We propose means of restructuring naming practices so that underrepresented groups have greater agency and visibility. The tools, approaches, and methodologies offered by the authors of the following chapters can be adapted for many cultural and geographical milieux.

Notes

1 Abraham R. Matamanda et al., eds., *Urban Geography in Postcolonial Zimbabwe* (Cham: Springer, 2021); James Duminy, ed., *Planning and the Case Study Method in Africa: The Planner in Dirty Shoes* (New York: Palgrave Macmillan, 2014).

2 Lauren Beck, *Canada's Place Names and How to Change Them* (Montreal: Concordia University Press, 2022); Reuben Rose-Redwood et al., eds., *The Political Life of Urban Streetscapes: Naming, Politics, and Place* (New York: Routledge, 2020); James Akerman, ed., *Decolonizing the Map: Cartography from Colony to Nation* (Chicago: University of Chicago Press, 2017).

1 Geotags and Check-Ins as Renaming Practices of Indigenous Digital Activism

The Gidmt'en Checkpoint Case

Anna Mongibello

The present study seeks to advance an understanding of geotagging and check-ins as reappropriation strategies, decolonial and counter-discursive resistance practices to the erasure of Indigenous place names in Canada and the restoration of Indigenous sovereignty over their lands. When used for political reasons, these practices can be seen as forms of digital social action helping to shape the reality of the Gidimt'en checkpoint on Facebook, thus amplifying the resonance of the Wet'suwet'en resistance in British Columbia and beyond. The chapter employs a Corpus-based Social Media Critical Discourse Analysis approach to identify how alternative narratives of place can circulate through social media platforms and potentially work as decolonizing tools for reclaiming ancestral names. In order to do so, it examines patterns of information flow focusing, in particular, on the combined use of geotags, check-ins, and hashtags to identify attitudes and discourses, and the way Indigenous activists reclaim their right to name unceded land.

The Wet'suwet'en territory is a 22,000 km^2 land in northwest British Columbia (Canada) that the Wet'suwet'en people, a governing nation that has existed since time immemorial, has inhabited for thousands of years. The territories have never been ceded to the Canadian government, which means that the Wet'suwet'en have never signed any treaty or agreement to surrender their sovereignty.[1] In 1997, with the *Delgamuukw vs. British Columbia*, the Supreme Court of Canada ruled that Aboriginal title is not extinguished in the areas claimed by the Wet'suwet'en. The action, launched in 1984 by the Gitxsan, another Indigenous nation, and Wet'suwet'en hereditary chiefs, saw ten years of litigation during which the Wet'suwet'en and Gitxsan had to prove their legal land claims according to the requirements and standards of the Canadian legal system.[2] The request was particularly challenging, since Indigenous land rights rely on oral storytelling, which has always been dismissed by Canadian legal and political authorities as mere cultural artifacts having no legal status. In the *Delgamuukw vs. British Columbia* case, however, the Court recognized that Indigenous oral histories provided evidence of ancestral practices and rightfully supported that Aboriginal title had not been extinguished when Canada became a country.

DOI: 10.4324/9781003414773-2

Despite having won a landmark case, since 2007, the Wet'suwet'en have strongly opposed the occupation of their lands by the federal and provincial governments of Canada and British Columbia backing oil corporations' requests to build pipelines throughout sacred unceded territory. One of these pipeline projects is the Coastal GasLink. The project consists in the construction of a 670-km-long natural gas pipeline that would carry fracked liquefied natural gas to a terminal in Kitimat (B.C.) for export to global markets. In 2007, the Wet'suwet'en expressed their will to prevent pipeline constructions through their territory. For this reason, in 2009 several checkpoints and resistance camps were established at the entrance to the Wet'suwet'en territory so as to oppose land occupation. When, on November 23, 2018, the Supreme Court of B.C. granted an injunction filed by Coastal GasLink to prevent Indigenous protesters from blocking the access to their own lands and the construction works toward the sacred river Wedzin Kwa, the Gidimt'en checkpoint was built at 44.5 km of the Wedzin Kwa. The checkpoint is directly in the path of the Coastal GasLink pipeline; it includes a fortified gate and a blockage controlling access to the only bridge entering unceded territory. The checkpoint is named after the Gidimt'en, one of the five clans of the Wet'suwet'en Nation. Since it was established, the checkpoint has been a resistance spot and a permanent camp for the Wet'suwet'en land defenders to contrast the violent invasion of their territory. It has also hosted gatherings, workshops, and traditional activities. For this reason, the checkpoint has soon become a symbol of the Wet'suwet'en resistance to the Coastal GasLink project.

Regardless of its increasing political and symbolic relevance, the checkpoint does not exist on official maps, although it stands, *de facto*, as a reclaimed piece of land and an outpost of resistance to those who join the protest. On January 7, 2019, the Royal Canadian Mounted Police conducted a militarized raid against the land protectors at the Gidimt'en checkpoint. A convoy of 35 RCMP vehicles with 60 to 100 militarized police troops attacked the protesters with assault weapons. Images of assault rifles pointed at unarmed Wet'suwet'en women circulated rapidly throughout social media platforms. The intervention saw the arrest of more than 10 Wet'suwet'en land defenders. This is when the protest saw a digital turn. An official account of the Gidimt'en checkpoint was opened on Twitter, generating an outpour of support and solidarity for the Wet'suwet'en. Then, few months later, in March 2019, the Gidimt'en checkpoint was added as a custom location on Facebook as "Gidimt'en Checkpoint—44.5," thus allowing people to check in there, locally and remotely.[3] The number (44.5) indicates its exact location, using the sacred Wedzin Kwa as a referent point to its physical location. Wet'suwet'en land defenders and allies have therefore started to use geotagging and check-in tools on social media in order to geotag themselves to the checkpoint. A total of 104 people checked in by October 2022, while 151 Facebook users are reported to be following the location page.

The Politics of Geotagging: Indigenous Place Names and Sovereignty

The power of place names has been traditionally explored across a variety of disciplines in the humanities and social sciences, ranging from literary studies, anthropology, and philosophy to geography, cartography, and linguistics. A common acknowledgment is that naming a place is always "a socially embedded act," one that most certainly involves uneven relations of power.[4] This is because names in general, and place names in particular, are not just meaningless marks that conveniently denote geographical spaces. On the contrary, place names have profound connotative meanings and therefore powerful discursive potential. As a matter of fact, place names *make meaning* and shape social relations. In other words, place names extend way beyond the mere linguistic realm, provided that we could really dismiss as being "merely linguistic" anything that has to do with language. As language is deeply political, place names are embedded in the social, cultural, and political dynamics of creating worlds and realities and, at the same time, of establishing power relations. In fact, place names seem to have a quintessential discursive quality due to the fact that they reflect and circulate ideologies. Place names enforced by British and French colonizers in Canada, for example, imposed a symbolic order that legitimized the dominant colonial ideology and went hand in hand with land occupation, as a transition was made to a new regime of ownership.[5] Therefore, a perspective that we embrace in this chapter is one that foregrounds the intersection of place-naming and power, as well as place-naming and resistance to power.

As a form of knowledge production, place-naming has been central in the enforcement of "centralized representations of the social world at the expense of the lower classes and speakers of non-sanctioned languages and dialects."[6] This is even more true when it comes to countries that experienced the violence of colonization like Canada, where the systematization of place names involved the consolidation of colonial authority. As a matter of fact, the imposition of English and French toponyms erased the heteroglossia of Indigenous place names. As a consequence, the history of Indigenous peoples was removed, while the myth of *terra nullius* was discursively created through place names. Suffice it to think of place names like Newfoundland or Nova Scotia that embed the narrative of the virgin land to be discovered and conquered.

On the other hand, Indigenous lands were never seen as owned by the Indigenous peoples living therein. On the contrary, the lie of *terra nullius* justified not only the physical but also the symbolic occupation of the land through its violent naming. At the same time, French and British place names reflected the ideological standpoint of the colonizers in that they embedded the history of European colonization. A common practice was that of naming places after people, therefore advancing an understanding of the land as a space to be conquered and owned. Quite the reverse, Indigenous place names had (and still have) a cognitive function since they carry ancestral stories and traditional knowledge. They also reveal land uses and express Indigenous land rights and sovereignty. In recent years, Indigenous peoples have been reclaiming their ancestral place names as part of their decolonization struggle and revitalization processes. The archipelago Queen Charlotte Islands, just to give an example, was named after Queen Charlotte by Captain George Dixon in 1787, thus erasing the ancestral name Haida Gwaii, which is how the Haida Nation used to call the islands. However, the traditional name was legally restored in 2010, when the province of B.C. passed Bill 18, the Haida Gwaii Reconciliation Act.[7] In 2019, the Geographical Names Board of Canada, a national body comprising representatives of provinces and territories, developed an interactive digital map of Canadian place names of Indigenous origins.[8] The map shows how Indigenous place names have evolved over time. While many Indigenous place names have been restored into the official map of Canada, much work is still to be done in this regard. Digital technologies are now offering new tools and spaces for reclaiming Indigenous sovereignty over ancestral lands, also through the restoration of and the struggle for Indigenous names.

Place Names in the Digital Era

In recent years, the digital tools offered by digital and social media have radically changed how we sense space. Geosocial media, or location-based social networks, like Foursquare and Yelp, "reflect a user's geographical action in the real world."[9] Thanks to these services, users can check in at real-world locations, such as restaurants, parks, stores, etc., in a given area and share their experience with other users. Previous research has shown that location information is used to recommend points of interest, provide directions, or facilitate social coordination among groups of friends.[10] Through comments and personal contributions, user-generated content exploiting geolocational digital resources can even affect the way we perceive places.[11] As a matter of fact, these contents create narratives about places that are de-centralized and structure the way we get to know or experience those places.

Social media platforms have also become popular sites for sharing geographic information. Hashtags on Twitter, for example, usually mark conversations by saying what a statement is really about.[12] In this sense, hashtags work as social practice in that they aggregate and enable the circulation of discourses around specific topics, which may also inscribe evaluative positions of the user, expressing attitudes and emotions. Although their topic function has been thoroughly explored by scholars,[13] hashtags can also aggregate information or anchor conversations to physical places. This is particularly common during social protests when users use location-based hashtags to virtually join the protest even if they are not physically present where the action is taking place. Similarly, other platforms like Facebook and Instagram allow the labeling of virtual content with geographic coordinates. Facebook, for instance, introduced the check-in feature in 2010. Since then, it has been a popular tool used by Facebook users to share their location. In the beginning, the feature was used for advertising purposes by businesses adding their location on Facebook so that customers could check in and share the business's location with their friends. Now users can also add location data to photos or posts that are shared in real time. Geotagged posts, photos, and check-ins provide information about the presumed location of the user. These digital contents contain spatial information or geospatial metadata through an indexing mechanism; users can, in fact, signal where a particular activity is taking place or simply their physical presence in specific geographic spots.

Considering the above-mentioned uses of geographical information embedded in social media content, scholars have postulated the so-called localness assumption,[14] based on which we can assume that a person posting a geotagged tweet is doing so from the place corresponding to the geotag. Usually, the user would type in or select their location from a list of places already existing in the platform database as they appear on official maps. However, geotags can also work as a complex mechanism for the digital co-production of place, the manipulation of locational data, and as practices of resistance, in general. In fact, users can either select from a pre-existing list of places or add their location, thus creating new place names. The latter seems to be a particularly prolific strategy regarding (post-)colonial settings.

The subversive potential of geotagging features as counter-discursive tools in the digital era can only be fully acknowledged in relation to an understanding of the metaverse as a sociopolitical environment.[15] According to Indigenous theorists of cyberspace, the virtual realm is not less "real" than the physical one, although the Internet has often been represented as landless. On the contrary, as Pascua Yaqui/Chicanx scholar Marisa Duarte explains, the Internet requires land-based infrastructures, and therefore, the relationship between the digital realm and the land is inextricable. Such a relationship is further complicated when it comes to cyber-corporations expropriating Indigenous lands to set up their towers, thus unsettling Indigenous rights to sovereignty and self-determination. At the same time, the digital realm can be seen as a reenactment of western cultural consciousness or a western product replicating the hegemonies of whiteness. In other words, virtuality could be understood as a new *terra nullius* conquered and controlled by non-Indigenous peoples, where the violence of settler colonialism is still perpetuated. It becomes more apparent, then, that the politics of geotagging can be particularly subversive if considering cyberspace as an extension of colonial settings. Geotagging offers another possibility for enacting decolonization practices and reshaping land-based relationships. This is because geotags allow Indigenous and ally users to participate actively in the semiotic process of spatial production.

In the last ten years, at least, Indigenous peoples have appropriated the cybersphere and used digital devices in ways that are creative, powerful, and political, thus facilitating new modes of resisting colonial hegemonies. By engaging online through social media such as Facebook, Instagram, Twitter (now X) and Tik Tok, Indigenous peoples "address the practices of de-legitimation and invisibilization, push back against existing forms of representation and ultimately advance a decolonial agenda."[16] The digital realm serves as a locus of resistance, offering opportunities to advocate for Indigenous rights. Indigenous users are actively re-shaping cyberspace, reclaiming spaces for discourse and mobilization. Movements like #idlenomore, #SOSBLAKAUSTRALIA, #IndigenousDads, and #NoDAPL exemplify Indigenous resistance on social media platforms, confronting threats to land rights, cultural erasure, and harmful stereotypes. Through these digital channels, Indigenous peoples assert their agency and demand recognition in the face of ongoing colonial injustices.

Previous Research

In recent years, there have been several studies exploring the potential and the practice of geotagging in network services.[17] While some research work has focused on web search data and geographic information,[18] the use and role of geotagging in digital activism have opened up a new line of research, mainly thanks to the increasing popularity of social media platforms and online protests in the last few years. Scholars investigating the use of geolocation data for political causes against perceived data surveillance have explored this latter line of research. An example is provided by Elson et al., who analyzed the 2009 protests in Iran when protesters invited Twitter users to geotag themselves in Tehran to confuse local authorities; in so doing, the people who were actually involved in the physical protesting were protected.[19] Another example is that of the Vinegar protests in Brazil, in 2013, when users tweeting about the protest checked into places where they were not physically located to support the movement remotely. Bastos et al. observed that X (Twitter) users disproportionally focused their attention on the southeastern region of Brazil and that this attention was not consistent with the distribution of protesters attending the demonstrations.[20]

Not much critical attention has been devoted to how Indigenous peoples have used geotagging features as a medium for reclaiming the right to place naming or place production. Karsgaard et al. investigated how Indigenous users make use of the free-form location profile field on Twitter to situate themselves using Indigenous place names that do not appear on official maps.[21] In fact, as the authors explain, Twitter's open location field enables users to place themselves through anti-colonial positionings. In so doing, they resist the tyranny of colonial place names by choosing to re-instate Indigenous toponyms, also in Indigenous ancestral languages, in ways that are subversive and empowering. A study by Baik examined how Indigenous protesters and allies of the #NoDAPL movement used geotagging and check-in features in 2016 to create the digital space of the protest. The users geotagged themselves at Standing Rock from remote places in an attempt to confuse the police's alleged surveilling activities.[22] The author advances the idea that check-ins potentially create a geotagging counter-public in the context of online activism.

These studies and the cases they consider attest to the subversive quality of geotagging features. When these features are employed by Indigenous and ally users, geotags can advance toponymic struggles in a variety of forms. In general, the counter-discursive potential of geotagging features has been overlooked by researchers working in the sociolinguistic analysis of digital protests and by technology studies scholars. Therefore, in light of the existing literature on the topic, this chapter aims to contribute to the ongoing research on the politics and practices of geotagging enacted by Indigenous and ally users. It advances an investigation of geotagging counter-discursive potential through qualitative and quantitative analysis. More specifically, we focus on how geotags and check-ins have been used during the Wet'suwet'en protest on Facebook as part of the digital resistance to the Coastal GasLink pipeline project. The analysis aims to address the following research questions:

RQ1: How do geotags and check-ins contribute to the discursive construction of the Gidimt'en checkpoint?
RQ2: What are the main discursive strategies employed by Wet'suwet'en and ally users when geotagging themselves at the Gidimt'en checkpoint?
RQ3: What are the implications of and the role played by geotags and check-ins in the wider struggle for the decolonization of Indigenous place names?

Data and Methodology

In order to address the research questions, a corpus of Facebook check-in and geotagged posts to the Gidimt'en checkpoint—44.5 was collected. The term "Gidimt'en" was entered on the Facebook search bar to retrieve relevant posts. The search yielded 52 Facebook posts by users who geotagged themselves at the Gidimt'en checkpoint. All publicly available posts geotagged to the Gidimt'en checkpoint were downloaded manually. The data span from March 2019, when the first geotagged post was published, to October 2022, when data collection ended. Content was reviewed in a plain text editor and embedded encoding was removed. While many posts contained both textual and visual content (i.e., photos), only text-based metadata were collected for the purpose of the study. Each file was saved in UTF-8-encoded plain text format. The final dataset, the GEO_GTEN corpus, encompasses 4,129 words and 3,488 tokens. The GEO_GTEN is a small, specialized corpus, thus allowing us to gain insight into the patterns of language use pertaining to a particular setting, that is, Facebook check-in and geotagged posts. Given its small size, our corpus guarantees a reliable representation of how Indigenous and ally users construe the Gidimt'en checkpoint virtually. Since all geotagged posts from the Gidimt'en checkpoint—44.5 location page were retrieved, the GEO-GTEN is representative of the usage patterns and online population under investigation, in relation to a specific socio-political event, that is, the Wet'suwet'en protest. It is therefore suitable for the purpose of the present research.

Corpus tools were used to complement Critical Discourse Analysis (hereafter CDA).[23] The combination with a CDA-informed approach is particularly fruitful to interpret the ideological implications in language use. Such an integrated approach, which combines qualitative readings with a corpus-linguistics approach, creates a methodological synergy[24] that has proven to be a reliable method to track down discursive patterns. The synergy between qualitative and quantitative methodologies addresses the limitations of both. In other words, "by using computer software, analysts can deal with much larger quantities of data and so put forward more convincing evidence in support of their claims."[25] Moreover, in embracing this methodological perspective to conduct our analysis, we agree with KhosraviNik[26] that we cannot see social media merely as a data source. On the contrary, as Bouvier maintains, social media data are not about a virtual realm, but they tell us about actual people and what they do with their online writing.[27] This is why understanding and acknowledging the context of production is essential so as to also understand the kind of discourses and the sociopolitical interests connected to those texts.

For the quantitative informed analysis, the online software program *Sketch Engine*[28] was employed as a corpus manager and text analysis software. We first retrieved a frequency list recording the occurrences of each word in the corpus. The frequency list was useful in order to identify and code recurring themes in the GEO_GTEN based on the most ubiquitous words. Frequency lists are a good starting point for the analysis of any type of corpus, in that they provide a sociological profile of a given word in its context of usage, and they can also direct the research in its initial stages. The Brexit Corpus was chosen as a reference corpus for comparing frequency lists. While we could have opted for a keyword analysis in order to spot salient words in the GEO_GTEN, given the dimension of the focus corpus, we expected the data to be misleading as a keyword list would have simply showed that the GEO_GTEN is a specialized corpus. On the other hand, we found the comparative analysis of frequency lists to be particularly useful in order to see whether there are any significant similarities or differences between the most recurring items in the language of geotags compared to a larger corpus of tweets. Although the Brexit Corpus is not exclusively a corpus of social media data, however, 95.6% of the text documents are Twitter posts, which explains why we decided to use it as a reference corpus, since the genre is similar.

While frequency lists are helpful in that they reveal the focus of a corpus or a text, a concordance analysis allows a closer examination of the occurrences of a search term in the context it occurs. At the same time, concordances reveal patterns between words, phrases, or lemmas that suggest the creation of a (counter-) discourse. Baker explains these patterns in terms of discourse prosody, that is, the relationship of a word to speakers and hearers in ways that reveal attitudes and evaluations.[29] The analysis of concordances was conducted in combination with collocational analysis, that is, the analysis of associations between words. Concordance and collocational analyses served as a gateway to CDA, providing indications of discourse prosody. Discourse prosody indexes the topics and issues associated with a given keyword. Both analyses were helpful in the detection of narratives of place related to the Gidimt'en checkpoint

The analysis also draws from Social Media Research methodologies as we aimed at investigating how semiotic resources like hashtags are used in combination with geotags and check-ins to the Gidimt'en checkpoint—44.5. Hashtags have recently attracted much critical attention as new communicative affordances. They have been defined as a unique form of user-generated metadata[30] or conversational tagging. The use of the # symbol marks conversations on social media, by saying what a statement is really about.[31] At the same time, hashtags can also contain locational clues regarding where an event is taking place.

Results and Discussion

The first part of the analysis consists in the identification of recurrent themes in the GEO_GTEN. Using the corpus analysis software *Sketch Engine*, we retrieved and compared two frequency lists, one pertaining to the most frequent words in the focus corpus, and another one encompassing the most frequent words in the reference corpus. Although the two corpora are different in size, the Frequency per Million (FPM) score allowed us to consider rates of occurrence and was therefore used as a standardized value. The complete list contained 177 items, but, for the purpose of the analysis, we set a cut-off point so that only the words that showed higher frequency per million rates compared to their frequency per million rates in the reference corpus were included. Our final list consisted of 70 items that are specific to the way Facebook users linguistically construe the Gidimt'en checkpoint—44.5 in our corpus. All linguistic items were divided into categories based on their meaning and function. We therefore grouped together the words pertaining to the same or similar semantic field. In so doing we obtained three categories encompassing words pertaining to the field of location and movement, personal references and time, as Table 1.1 shows. For space restraints, we will only consider the first two categories.

The categories in Table 1.1 can be interpreted as relevant themes in the GEO_GTEN. The category of location and movement encompasses more terms

Table 1.1 Items grouped by semantic category.

Items	*Category*
here, through, under, around, up, across, go, come, place, where, down, at, there, from, way	LOCATION AND MOVEMENT
my, her, me, our, your, she, member, they, them, people, us, we, you	PERSONAL REFERENCES
today, time, since, still, while, future, week, long, day, when	TIME

Table 1.2 Collocates of *here* in the GEO_GTEN corpus.

Word	*Co-occurrences*	*Candidates*	*MI*
we	10	14	8.62
on	5	44	4.70
are	3	24	4.84
is	4	49	4.23

compared to the other two. The preposition *here* shows a much higher FPM rate compared to the reference corpus (4,359.41 as opposed to 730.89). The same difference does not apply to another deictic term, *there*, although the FPM rate is still higher in the focus corpus (2,664.08 vs. 1,835.35). The difference in the usage of deictics in the reference and the focus corpus means that more emphasis is placed on location information in the GEO_GTEN. This is a relevant yet expected result. In addition, the deictic *here* signifies that, when geotagging themselves at the Gidimt'en checkpoint—44.5, Facebook users are also physically present there and therefore comply with the localness assumption. We could assume that whenever the deictic term *here* occurs, the geotags can be classified as locative in that geotags and check-ins are used to locate the user physically. We further investigated *here* by retrieving its collocates to check whether any patterns occur. The word span was set to 4L/4R in the collocate window. The minimum frequency of a collocate to co-occur with the node was set to 3.

Table 1.2 shows that the collocate with the highest MI score is *we*, followed by a preposition (*on*) and two verbs that may indicate an existential process (*are* and *is*). *Here* is, therefore, locative in that it is used in the GEO_GTEN to locate the user where the post is being geotagged and, at the same time, a collective identity is embraced. The collective identity serves to the identification of the self with a group of people sharing the same location and the same purpose. Presumably, the collective *we* indicates the existence of an in-group made of

Wet'suwet'en protesters and their allies who are resisting at the checkpoint, as the following post (Example 1) shows:

We have always been here. Our ancestors tanned hides here—camped around the river and the creek area. We utilized the creek as our life source (water), and even to soak our hides.
We are still here. We tanned hides here today—camped around the river and creek area. We utilize and protect the creek—as it is our life source.
We will always be here. We will continue to protect our life source—as it sustains us, our neighbours, and generations to come.
#wetsuwetenstrong #notrespass #wedzinkwa #noconsent #nopipelines #noproblems #waterislife #protecttheyintah #landdefenders #waterprotectors #unitethenations #indigneousrising

The post shows that, by referring to the in-group (*we*), the user locates him/herself in time and space as the deictic *here* is used in combination also with items pertaining to the category of time (*always*, *still*). The Gidimt'en checkpoint—44.5 is therefore referred to as a physical place where a collective identity is reinforced. Timeliness emerges as a legitimation strategy to justify the physical presence of the user and the in-group protesting against RCMP on the Wet'suwet'en land. While the post is published on Facebook, the deictic *here* refers to the materiality of the checkpoint, thus establishing a continuity between the digital and the physical realm. The Gidimt'en checkpoint—44.5 is claimed and shaped as a place of resistance where the protection of the land takes place. The verb *protect*, repeated twice, in combination with hashtags like #landdefenders, #waterprotectors and #notrespass, echoes a discourse of war. These hashtags in particular are representative of the discourse of the Wet'suwet'en protest,[32] encompassing themes like mobilization (#protecttheyintah, #unitethenations, #noconsent, #nopipelines, #indigneousrising), land claim, and Indigenous sovereignty (#wetsuwetenstrong, #landdefenders, #waterprotectors). In other words, when geotagging themselves at the checkpoint, the users do not simply share where they are physically located, but they also construe the place that otherwise would not exist on official maps. The combination of hashtags and geotags embeds the narrative of place within the broader digital conversation on the Wet'suwet'en protest. Through geotags and check-ins, users seem to join a collective identity and a collective action, while also reinforcing the message that a collective community exists at the checkpoint. The existence of a collective community legitimizes the existence of the Gidimt'en checkpoint—44.5 as a real place.

Based on preliminary data, we can hypothesize that the digital construal of the Gidimt'en checkpoint—44.5 is very much connected with issues concerning identity affirmation and self-definition. Interestingly, personal pronouns and adjectives all register higher rates of FPM in the focus corpus, compared to the reference corpus, which we may read as signaling an overuse of personal references in the GEO_GTEN compared to the Brexit Corpus. These encompass personal references in the singular and plural forms, further identifiable as individual and collective identity markers. Individual markers include *my, me, her, she*; collective markers include *they, them, our, us, we*. Uses of *you* fall in both categories depending on the context. Among the individual markers, *my* occupies the highest ranks in the focus corpus; therefore, we decided to further investigate its concordances and collocates to see how it relates to the construction of the checkpoint.

The strongest candidates of *my* based on the MI score are *children*, *ground*, and *standing* (Table 1.3). By expanding the search to check the KWIC, we observed a repetition of the expression "standing my ground for," where the term ground indicates the position that the user will occupy soon at the Gidimt'en checkpoint—44.5 therein geotagged, as the following post (Example 2) shows: Konnorónkhwa Riken'ó ❤ "I'm standing my ground for my children and I'm standing my ground for your children."—Logan Staats I'll be standing beside you soon. Packing my bag and getting ready to protect the water.

Here again, although the reference is personal, the context highlights that the Gidimt'en checkpoint—44.5 is construed as a place of action, in this case the protection of sacred water. Compared to the use of the geotag in Example 1, the geotag in Example 2 is predictive of the user's intention to join the protest soon at the checkpoint. The use of *-ing* forms increases immediacy. Through direct address ("I'll be standing beside *you* soon") the user refers to the community of people that is already located at the checkpoint and that physically exists beyond the digital realm. The code-switch from Mohawk/Kanyen'kéha in the opening sentence ("Konnorónkhwa Riken'ó"— "I love you little brother") to English is another identity marker that contributes to the construction of the user's identity in discourse. If personal and collective references to an in-group are frequently used in the GEO_GTEN in locative geotagged posts in combination with the deictic *here*, or in predictive geotags, at the same time uses of *they* and *them* suggest that an out-group may also be construed in geotagged posts. In order to check this hypothesis, we investigated concordances and collocates of *they*, being the collective pronoun potentially referring to an out-group with the highest FPM rate in the focus corpus.

Table 1.3 Collocates of *my* in the GEO_GTEN.

Word	*Co-occurrences*	*Candidates*	*MI*
children	3	5	6.75
ground	3	6	6.49
standing	3	12	5.49

Table 1.4 Collocates of *they* in the GEO_GTEN.

Word	*Co-occurrences*	*Candidates*	*MI*
when	3	6	6.37
are	4	24	4.78
Wet'suwet'en	3	20	4.63
RCMP	4	28	4.56

Table 1.4 shows that *they* alternatively refers to RCMP (4) and the Wet'suwet'en (3). Our hypothesis of an out-group construction is, therefore, only partially confirmed. In Example 3, for instance, *they* refers to the Wet'suwet'en. The geotagged post advances a narrative of Indigenous life at the camp interrupted by RCMP:

> These two Wet'suewet'en elders live at Gidimt'en camp, "Aunty" Janet & Lawrence. *They*'re subject to RCMP barging onto their lands multiple times a day. The unwanted CGL pipeline's under construction nearby. Only two other land defenders *there*. They sure look threatening, don't they?

In this post, the user geotags him/herself at the Gidimt'en checkpoint—44.5, although he/she is not physically present there. The use of the deictic *there* suggests, in fact, a remote use of geotagging. As outlined in previous research, remote geotagging has often been used in the context of online activism for showing remote solidarity. In this specific case, solidarity is shaped by means of informal nomination combined with instances of relational identification and classification ("aunty Janet & Lawrence"—"Wet'suwet'en elders"). In fact, the way the social actors are represented shows the author's alignment (i.e., solidarity) with the Wet'suwet'en positions. By representing Janet as "aunty," and also referring to her and Lawrence as "elders," the user recognizes their authority among the Wet'suwet'en (since Indigenous elders play an important spiritual leading role) and the matriarchal role of Janet, specifically, within her community at the checkpoint. The discourse of solidarity is also conveyed through word choices like *barging* and *unwanted*, portraying the RCMP and Coastal GasLink's actions as violent and illegal, as they violate unceded lands. Therefore, solidarity with the Wet'suwet'en is presented within a cause-effect framework: connecting, even remotely, with the Gidimt'en checkpoint is the logic response to the violent threat the Wet'suwe'ten are undergoing. So much so that the post ends with a statement and a question tag embedding a rhetorical question. Through the rhetorical question, the user invites potential receivers of the message to verify for themselves, by looking at the photo attached to the post, whether the elders are threatening to the point that an armed force intervention is required to stop them. One last point worth mentioning is how the Gidimt'en checkpoint—44.5

is referred to as a camp where Wet'suwet'en elders and other land defenders live. The discourse of solidarity intertwines with the representation of Indigenous social actors. Emphasis is put on the material process of *living* on camp, which further advances the idea that the Gidimt'en checkpoint is a physical space where people exist and resist.

A discourse of solidarity is also present in Example 4, where it combines with a call for action. *They* refers, alternatively, to RCMP patrolling Wet'suwet'en land and to the Wet'suwet'en living on their own land. The user geotagging him/herself at the checkpoint denounces RCMP's assaults, therefore expecting the rest of the virtual community to sympathize with their feelings and align with them. Deictics like *now* (repeated twice) and *this* reinforce the connection in time and space between the user, his/her message, and his/her physical or remote presence at the checkpoint. In so doing, the user seems to be reporting from the Gidimt'en checkpoint. It could also be argued that such deictics may create an illusion of physical presence, whereas the user is not actually located where the post is geotagged. However, the illusion is achievable because geotags allow both locative and remote usages:

> @rcmpgrcpolice Helicopters are now circling Molly's house where there is Children and Babies live!!!! This is soooo Fucked up! They live on the Land and this is crossing the line even though that line was crossed years ago! This is a clear threat to those Indigenous Children and Babies!!! I am enraged and you should be too!!! Now their [sic: they're] going to raid Molly's home and point their high powered rifles and small Indigenous Babies!!!! 😡😡😡😡😡 RCMP CGL NDP GET THE FUCK OFF THE YINTAH!!! #alloutforwedzinkwa #wetsuwetenstrong #rbciskillingme

Devices like mentions (@rcmpgrcpolice), emoji (😡😡😡😡😡), hashtags (#alloutforwedzinkwa #wetsuwetenstrong #rbciskillingme), and capitalized letters express the urgency associated with a negative situation (helicopters circling Molly's house and rifles pointed at Indigenous children) and the necessity for quick actions to be taken. Here the geotag serves a double function: on the one hand, it locates the narrative within the framework of war and resistance at the Gidimt'en checkpoint; at the same time, it is also meant to socially engage other readers of the post, symbolically, emotionally, or physically, in the resistance taking place at the camp, since the geotag allows the user to publicly expose his/her own location to a broader audience. Therefore, the geotag foregrounds the social situation of the Gidimt'en checkpoint while also framing the checkpoint not merely as a physical space but also as a place having deep social meaning. By sharing locational information, the user is affiliating emotionally with the battle of the Wet'suwet'en and is publicly displaying his/her own engagement, hoping to raise similar responses among a wider Facebook audience. Overall, the use of the geotag in Example 4 seems to be affiliative as it aims at communicating the

Table 1.5 Collocates of *Gidimt'en* in the GEO_GTEN.

Word	*Co-occurrences*	*Candidates*	*MI*
checkpoint	10	10	7.43
camp	6	14	6.20
Wet'suwet'en	6	20	5.69

user's affiliation to the battle and, therefore, it works as an endorsement showing personal support.

The data retrieved so far have shed some light on the discursive strategies of Facebook geotagged posts and check-ins to the Gidimt'en checkpoint. However, one missing piece in the analysis is how the Gidimt'en checkpoint is construed in the GEO_GTEN when the keyword *Gidimt'en* occurs in the posts. Therefore, we decided to investigate *Gidimt'en* in the GEO_GTEN by looking at its collocates and concordances.

Table 1.5 shows that the collocate with the highest MI score is *checkpoint*, followed by *camp*. The data confirm that *Gidimt'en checkpoint* is the name that Facebook users prefer when referring to the Wet'suwet'en resistance and related activities in their posts. It is therefore both the place name assigned to that specific portion of territory on Wet'suwet'en unceded land and a powerful cultural signifier imbued with meanings related to resistance and sovereignty. As a matter of fact, when *Gidimt'en checkpoint* or, alternatively, *Gidimt'en camp*, is included in the post, references are made by users to colonization/colonial violence (lines 7, 8, and 15) or police harassment (lines 2, 4, 5, 14, 17), as Example 5 shows:

> Today, April 18th, 2022- the RCMP have mistakenly identified and arrested a supporter after a month of police harassment at the **Gidimt'en Checkpoint**. There has been over 100 instances of police trespassing at our village site in the last month with police frequently trespassing 4–5 times per day. This tax-payer funded harassment and intimidation is an explicit attempt to make Wet'suwet'en people unsafe on our own lands. Our supporters have been working tirelessly standing on the frontlines to protect our elders and community members from this police violence at our village site. We call on our supporters to come stand with us and get more boots on the ground.

This post was shared by an activist, allegedly posting from the checkpoint, as the last part of the text seems to suggest ("come stand with us"). Once again, the checkpoint is framed within a war-like scenario activated by words like *harassment, trespassing, frontlines, protect,* and *violence*. Within this framework, an opposition emerges between the supporters protecting Wet'suwet'en elders and community members, and the police harassing and trespassing on their sites.

The Gidimt'en checkpoint is presented as the location where both the RCMP harassment and the Wet'suwet'en protection of their frontlines are taking place. The checkpoint is situated on Wet'suwet'en land ("our own lands," "our village site"), with no further locational information provided. The lack of additional locational information is due to the fact that the physical location of the Gidimt'en checkpoint and what the checkpoint is in general is shared knowledge among the potential readers of the post. Therefore, the author of the geotagged post can assume that the potential readers are already familiar with the place name. *Gidimt'en checkpoint* is, in fact, the place name shared among and validated by the community of Facebook users checking-in at the checkpoint, the Wet'suwet'en activists and their allies. Even though the name is absent from official maps, it is overall employed and recognized as the official, reclaimed name of the resistance outpost on Wet'suwet'en land, as it reflects Wet'suwet'en sovereignty over their territories.

Conclusion

Since 2007, the Wet'suwet'en people, an Indigenous nation living on unceded land in British Columbia, have opposed the Coastal GasLink pipeline project that would carry fracked liquefied natural gas running through their sacred lands. The Gidimt'en checkpoint was built as a resistance site for the Wet'suwet'en land defenders to contrast the violent invasion of their territory. The checkpoint does not exist on the official map of Canada, yet it has become a place of cultural and political significance since it is the symbol of the Wet'suwet'en resistance to the Coastal GasLink and RCMP invasion.

The present chapter has provided insights into how geotagging tools on Facebook are used by Wet'suwet'en activists and their allies to reclaim the right to name the land and to validate the existence of physical and symbolic outposts of resistance, like the Gidimt'en checkpoint. The analysis, conducted on a corpus of geotagged posts retrieved from Facebook and named the GEO_GTEN corpus, revealed how the Gidimt'en checkpoint—44.5 is being discursively produced by Facebook users checking-in either physically or remotely at its location. Addressing RQ1 and RQ2 (How do geotags and check-ins contribute to the discursive construction of the Gidimt'en checkpoint? And what are the main discursive strategies employed by Wet'suwet'en and ally users when geotagging themselves at the Gidimt'en checkpoint?), a comparative analysis between frequency lists retrieved from the GEO_GTEN and the Brexit Corpus (used as a reference corpus) revealed that three categories of words are statistically more frequent in our focus corpus, that is, words pertaining to the category of location and movement, personal references, and time. Preliminary results showed that there is an overuse of deictics in the focus corpus, based on FPM scores, which required further investigation to see how deictics contribute to the discursive construction of the Gidimt'en checkpoint. The qualitative analysis showed that uses of *here* are locative, therefore establishing a continuity between the digital and the physical realm. Uses of *there* are predictive of the user's intention to join the protest or signal remote geotagging. In all cases, deictics combined with collective or personal identity markers, within a war-like scenario that we found being frequently used by Facebook users, shape the Gidimt'en checkpoint as a place of action and resistance, where a community of elders, activists, and allies exists. Remote geotagging intertwines with a discourse of solidarity and seemed affiliative as it aims at communicating the affiliation of the user to the battle of the Wet'suwet'en, thus working as an endorsement showing personal support as well as legitimation.

Based on our results, geotagging features contribute to foregrounding the Gidimt'en checkpoint on Facebook not merely as a physical space but as a place having deep social meaning. The discursive construction of the Gidimt'en checkpoint in the analyzed geotagged posts shows that the place name is inscribed within a narrative of war, solidarity, and call for action. It is therefore discursively produced, then claimed and circulated, as both a physical place located 44.5 km from Wedzin Kwa, and as a socio-cultural spot where resistance and violence are happening. In all cases, through the geotagged posts, the Gidimt'en checkpoint is validated and legitimated, regardless of its absence on the map of Canada. Overall, the results prove that geotagging features can be valuable tools in the Indigenous struggle for claiming the right to name as they validate the existence of places of cultural and political importance to the Indigenous peoples. The data showed that geotagging and check-ins can work as reappropriation strategies and as counter-discursive resistance practices to both the erasure or even the invisibility of Indigenous place names and the restoration of Indigenous sovereignty over their lands.

An aspect that was not included in this research is how geotagging features combine with visual material in the semiotic construction of place. This is one limitation to this study that further research could investigate. Also, further research should be done in regard to how comments and likes of geotagged posts co-contribute to the legitimation of place and the co-validation of place names.

Notes

1 Pamela Palmater, *Warrior Life: Indigenous Resistance and Resurgence* (Halifax: Fernwood Publishing, 2020).
2 Val Napoleon, "Delgamuukw: A Legal Straightjacket for Oral Histories?" *Canadian Journal of Law and Society* 20 (2005), pp. 123–155.
3 https://www.facebook.com/pages/Gidimten-Checkpoint-445/387577721791680
4 D. L. Berg and J. Vuolteenaho, eds., *Critical Toponymies: The Contested Politics of Place Naming* (Burlington, VT: Ashgate, 2009), p. 10.
5 Anna Mongibello, "Place-Names and Self-Determination: Toponyms in Indigenous News Discourse," in *Shifting Toponymies. (Re)naming Places, (Re)shaping Identities*, eds. Luisa Caiazzo and I.M. Nick (Newcastle upon Tyne: Cambridge Scholars Publishing, 2020), p. 113.

6 Berg and Vuolteenaho, *Critical Toponymies*, p. 4.

7 https://archive.news.gov.bc.ca/releases/news_releases_2009–2013/2010prem0125–000719.htm

8 The project, launched on June 21, 2019, which is National Indigenous Peoples Day in Canada, is called "Stories from the Land: Indigenous Place Names in Canada" and is hosted by the Government of Canada website: https://maps.canada.ca/journal/content-en.html?lang=en&appid=0e585399e9474ccf932104a239d90652&appidalt=11756f2e3c454acdb214f950cf1e2f7d.

9 Fred Morstatter, Huiji Gao, and Huan Liu, "Discovering Location Information in Social Media," *Bulletin of the IEEE Computer Society Technical Committee on Data Engineering* 38.2 (2015), pp. 4–13.

10 Henriette Cramer, Mattias Rost, and Lars Erik Holmquist, "Performing A Check-in: Emerging Practices, Norms and 'Conflicts' in Location-sharing Using Foursquare," in *Proceedings of the 13th Conference on Human-Computer Interaction with Mobile Devices and Services, Mobile HCI 2011*, pp. 57–66.

11 Sébastien Caquard, "Cartography II: Collective Cartographies in the Social Media Era," *Progress in Human Geography* 38.1 (2013), pp. 1–10.

12 Yarimar Bonilla and Jonathan Rosa, "#Ferguson: Digital Protest, Hashtag Ethnography," *Journal of the American Ethnological Society* 42.1 (2015), pp. 4–17.

13 danah boyd, Scott Golder, and Gilad Lotan, "Tweet, Tweet, Retweet: Conversational Aspects of Retweeting on Twitter," *43rd Hawaii International Conference on System Sciences* (2010), pp. 1–10.

14 Isaac Johnson, Subhasree Sengupta, Johannes Schöning, and Brent Hecht, "The Geography and Importance of Localness in Geotagged Social Media," *Proceedings of the International Conference on Human Factors in Computing Systems CHI* (2016), pp. 515–526.

15 David Gaertner, "What's a Story Like You Doing in a Place Like This? Cyberspace and Indigenous Futurism," in *Novel Alliances: Allied Perspectives on Literature, Art, and New Media*, https://novelalliances.com/2015/03/23/whats-a-story-like-you-doing-in-a-place-like-this-cyberspace-and-indigenous-futurism-in-neal-stephensons-snow-crash/.

16 Anna Mongibello and Bronwyn Carlson, "Indigenous Resistance in the Digital Age. An Introduction," *Anglistica* 25.1 (2021), p. 3.

17 Liangliang Cao, Gerald Friedland, and Pascal Kelm, "Second ACM Multimedia Workshop on Geotagging and Its Applications in Multimedia," in *Proceedings of the 21st ACM International Conference on Multimedia* (2013), pp. 1091–1092; Katharine Welsh, Derek France, Brian Whalley, and Julian Park, "Geotagging Photographs in Student Fieldwork," *Journal of Geography in Higher Education* 36.3 (2012), pp. 469–480.

18 Qingqing Gan, Josh Attenberg, Alexander Markowetz, and Torsten Suel, "Analysis of Geographic Queries in a Search Engine Log," *LOCWEB '08: Proceedings of the First International Workshop on Location and the Web*, (2008), pp. 49–56.

19 Sara Beth Elson, Douglas Yeung, Parisa Roshan, S. R. Bohandy, and Alireza Nader, *Using Social Media to Gauge Iranian Public Opinion and Mood after the 2009 Election* (Santa Monica: Rand, 2012).

20 Marco T. Bastos, Raquel de Cunha Recuero, and Gabriela de Silva Zago, "Taking Tweets to the Streets: A Spatial Analysis of the Vinegar Protests in Brazil," *First Monday* 19.3 (2014).

21 Carrie Karsgaard, Maggie MacDonald, and Michael Hockenhull, "Rename and Resist Colonial Extraction: Twitter's Toponymic Politics," *AoIR Selected Papers of the Internet Research* (2020).

22 Jeeyun Baik, "The Geotagging Counterpublic: The Case of Facebook Remote Check-Ins to Standing Rock," *International Journal of Communication* 14 (2020).

23 Lynne Flowerdew, "The Argument for Using English Specialized Corpora to Understand Academic and Professional Settings," in *Discourse in the Professions: Perspectives from Corpus Linguistics*, eds. Ulla Connor and Thomas Upton (Amsterdam: John Benjamins, 2004), pp. 11–33.
24 Paul Baker, Costas Gabrielatos, Majid KhosraviNik, Michal Kryzanowski, Tony McEnery, and Ruth Wodak, "A Useful Methodological Synergy? Combining Critical Discourse Analysis and Corpus Linguistics to Examine Discourses of Refugees and Asylum Seekers in the UK Press," *Discourse & Society* 19 (2008), pp. 273–306.
25 Deborah Cameron and Ivan Panović, *Working with Written Discourse* (Los Angeles: SAGE, 2014), p. 81.
26 Majid KhosraviNik, "Social Media Critical Discourse Studies," in *The Routledge Handbook of Critical Discourse Studies*, eds. John Flowerdew and John E. Richardson (New York: Routledge, 2017), p. 593.
27 Gwen Bouvier, "What Is a Discourse Approach to Twitter, Facebook, YouTube and Other Social Media: Connecting with Other Academic Fields?" *Journal of Multicultural Discourses* 10.2 (2015), pp. 149–162.
28 Adam Kilgarriff et al., "The Sketch Engine: Ten Years On," *Lexicography* 1 (2014), pp. 7–36.
29 Baker et al., eds., "A Useful Methodological Synergy."
30 Michele Zappavigna, *Searchable Talk. Hashtags and Social Media Metadiscourse* (London: Bloomsbury Academic, 2018).
31 Bonilla and Rosa, "#Ferguson: Digital Protest, Hashtag Ethnography."
32 Anna Mongibello, "#alleyesonwetsuweten: An Analysis of the Wet'suwet'en Protest on Twitter," *Anglistica* 25.1 (2022).

2 Decolonizing Place Names in Vietnam

An Overview of Names with Linguistic Diversity

Phùng Thị Thanh Lâm

The process of exploring, conquering, and exploiting new lands by colonists[1] imposed many new social elements on the land, called colonies by explorers. In the journeys of colonists, naming places, in other words, often assumed the lands were primitive areas over which they seized the right to name places and apply the names following conventions that they recognized. Yet, we know this colonial system of place names coexists with other practices created by native people, and that "the linkage between place names and power is more readily apparent in a contested and post-colonial setting."[2] This situation of place names in a post-colonial context frequently leads to a conflict between native people and settlers in that "the authorities had the power to select what were considered appropriate names [. . . to the streets of the city], but the [Asian] communities determined whether the names took on common usage."[3] From Indigenous people's view, losing lands by colonial invasion perforce causes them to refuse the colonial names of places because placed named in honor of colonialism are not associated with the "cultural identity of a people;" therefore, the "local people . . . do not know or understand what those non-indigenous place names mean."[4]

In the context of European colonization around the world, their naming of places located in colonized lands consists of two standard patterns: bearing names extracted from the colonizer's place of origin or transcribing native place names, often in the language and using the phonetic logic of the colonizer. In conquering new lands, place names became a valuable tool to honor those associated with the colonial process. Native names, however, often challenge the ears of colonists. In a study on Australian place names, Monaghan points out how, in the initial phase of naming during colonization, Indigenous place names were overlooked by colonists because they had little or no connection to the local inhabitants. Even if the colonists expected to use place names in the native language to name geographical features, this process was also tricky because the natives understood and spoke very little English, so it was difficult for both parties to exchange ideas. The transcription and appropriation of Indigenous place name knowledge then led to the introduction of innumerable erroneous place names associated with the process of misunderstanding place

DOI: 10.4324/9781003414773-3

names, sometimes in a humorous way, that did not reflect the salient features of geographical features.[5]

The colonists transcribed native place names based on the principle of phonetics and orthography of their mother tongues. Thus, Indigenous place names in Australia and some African countries were recorded with English influences whereas German influences are more prevalent in Namibia.[6] Even with the help of native people, the process of transcribing place names still faces difficulties; for example, pronouncing the retroflex and nasal sounds in English challenges Indigenous people in Australia. To reduce the difficulty of pronouncing African Indigenous place names, the colonists anglicized names alongside misnaming Indigenous African place names using English pronunciation, as well as naming places using nicknames, adding or omitting a letter in the native names, and simplifying a complex native name by using only the initial components of a complex name (e.g., Opi is the shortened name of Opinato). Similarly, in colonial Singapore, the "original names derived from Malay vernacular" had to be modified to "suit English-speaking tongues" even when they "were preserved;" to denote the attitudes of protestation with colonial place names, native Singaporean people "seldom comprehended and [were] frequently ignored."[7]

Decolonizing place names typically involves measures taken after the colonizer has left their names but themselves retreated from the land. In African and Asian countries under colonialism, renaming places was executed after their gaining independence, with the establishment of new governments in which colonial names were sometimes replaced by traditional names for local features, or local heroes and figures associated with the independence movement, as well as "activities that formed a significant role in daily experience."[8] From a nationalist perspective, renaming colonial place names could be considered an act of removing the pain of slavery and both reminding and stimulating within the people their patriotic spirit. Among the activities that resulted in renaming places, the municipal government in ex-colonial countries such as Vietnam and Namibia undertook a systematic renaming of colonial streets because those names honored colonists and glorified the victories of the invaders. Renaming streets is also judged to be an opportunity to close the door on the colonial era and in so doing, shut out the pain of the past so to look forward, instead, to a new bright future for an independent people.

Decolonizing place names therefore involves "a most potent strategy for restoring and preserving indigenous cartographies in post-colonial Africa" and "returning adulterated native toponyms to their original form."[9] However, sometimes, decolonizing place names comes up against the fact that local people feel attached to colonial names. In Vietnam, an important vehicle for popularizing its post-independence street names was a guidebook of street names launched by the government. It included a gloss configured as a column listing the new names it imposed next to another column listing colonial street names. This tool suggests that changing names on a larger scale can provoke challenges, especially at the local level where certain name types—such as those for streets—are so numerous that logistical challenges arise when so many change in a short period of time. Such was the case in Singapore, where "the colonial street names, in fact, were not removed as frequently as the names in other categories."[10]

If the renaming streets and removal of colonial street names after independence were profoundly analyzed by place names scholars, they would conclude that the subject still requires urgent and crucial research. Decolonizing place names recently involves promoting the opportunity to seek out place names in native languages through a diversity of pronunciation, emphasizing restoring some of the old names, as well as raising the question of authority in the naming of places.[11] In Vietnam, decolonizing place names involved two strategies: revising the inaccurate names of administrative units such as hamlets, villages, and provinces, as well as the names of geographical features in the early colonial period, and cleansing street names after the fall of the French colonial government in 1945 of references to colonial Vietnam.

This chapter focuses on the former aspect involving social and linguistic reasons for decolonizing place names in colonial Vietnam. In doing so, it will explore the role of the Vietnamese people in decolonizing and finding the political motivation behind the decolonization of place names in Vietnam during the colonial era. This chapter will focus on maps, plans, and cartographies of Hanoi city and its environs, a city that has been the capital of Vietnam for more than a thousand years and the capital of Indo-Chine under the French colonial period. The toponymy under study has been collected from maps held by the Bibliothèque nationale de France, with the scales of 1/10,000, 1/15,000, and 1/50,000 and dated from the early nineteenth century to 1945.[12] This date range reflects the period in which France colonized Vietnam, from 1883 to 1945. Furthermore, some maps published before 1883 are included in the analysis of place nomenclature because the names of the places written on those maps indicate an evolution in the system of writing place names in Vietnamese as a language. Place names denoted on the maps, including the names of regions, territories, districts, communes, hamlets, and villages, inform this chapter's analysis. In addition to the maps, this chapter researches these toponyms using numerous gazetteers and records about local areas that included place names written in Chinese and Nôm scripts published by Vietnamese governments.

Vietnamese Spoken Language, Multi-Scripts, and the Rise of *Chữ Quốc ngữ* (Vietnamese Romanized Script)

Vietnamese is a tonal, monosyllabic language that is spoken by the King ethnic group, which makes up 85.3% of Vietnam's population.[13] Scholars have established that the Vietnamese language evolved from the Vietnamese-Muong language group, a group of the Mon-Khmer family, starting in the eighth century.[14] Along the course of its history, Vietnamese coexisted with several scripts consisting of Chinese script (*Chữ Hán* in Vietnamese, 漢字), Nôm script (*Chữ Nôm* in Vietnamese, 𡨸 喃), French, and Vietnamese Romanized script (*Chữ Quốc ngữ* in Vietnamese*)*. Both leaders and intellectuals alike used Chinese script as an official and literary script from 111 BC to 1945 AD. Based on the Chinese script,[15] Vietnamese intellectuals created *Chữ Nôm* (𡨸 喃, Nôm script, ideographic script) that was only used in order to record Vietnamese language.[16] Unlike Chinese script, the *Nôm script* was used for individual purposes rather than as an official script. After European missionaries sailed to Vietnam to deliver sermons in the seventeenth century, they sought means to communicate directly with the local people. They used the alphabet of romance languages to record the sound of Vietnamese as a spoken language that then would be developed to become Vietnamese Romanized script *(Chữ Quốc ngữ)*. This Romanized script then allowed the European missionaries to remove Chinese script from political and social life because they believed that eliminating "the Chinese language was simultaneously [a means] to isolate Vietnam from its heritage and to neutralize the traditional elite."[17] In French colonial times (1883–1945), the colonial French government also installed French as an official language, with its related script, in every aspect of Vietnam's society; it also promoted Chinese script and Vietnamese Romanized script in French-Vietnamese schools.[18]

The French colonial government promoted teaching and learning *Chữ Quốc ngữ* based on the practical advantages of *Chữ Quốc ngữ*. This writing system uses Roman letters to display the sounds of Vietnamese and does not require a long time to learn. However, although Vietnamese students of the Collège des Interpretes could read and write *Chữ Quốc ngữ* quickly, they could not understand the lexical meanings of the words, especially because the terms were involving to have a new social life. A Vietnamese syllable may refer to some words that have different lexical meanings. Therefore, it is not easy to understand clearly without knowledge of Chinese.[19] The colonists recognized that without knowledge of Vietnamese history, culture, and social issues expressed by Chinese and Nôm scripts, students would not be able to interpret accurately. This situation led the colonists to conclude that "Chinese was still necessary for access to huge reservoir of knowledge . . . and that *quoc ngu* had considerable potential for mass education and proselytzing."[20]

As a practical measure, the colonists considered *Chữ Quốc ngữ* as a vehicle that could support them "to deal directly with the local population" as well as "a means to train minor native functionaries and to filter out incendiary ideas emanating from Paris."[21] Because of these political purposes, the colonial government promulgated policies forcing native people to learn *Chữ Quốc ngữ* by, for instance, teaching *Chữ Quốc ngữ* in schools, forcing the use of *Chữ Quốc ngữ* in administrative documents, and requiring Vietnamese bureaucrats to learn *Chữ Quốc ngữ,* if they expected to advance in their careers.[22] In 1869, the governor of Cochinchine Ohier promulgated a decree of applying *Chữ Quốc ngữ* in the composition of administrative documents. A few years later, in 1886, Résident Général Paulin Vial promulgated the decree to use *Chữ Quốc ngữ* in the enrolment exams of the Collège des Interpretes in Yên Phụ, Hanoi. The exams included writing an essay using *Chữ Quốc ngữ*, being able to read and write a simple essay in French, and being able to read, write, and translate Chinese, as well as knowing how to do mathematical operations, such as addition, subtraction, multiplication, and division.[23]

Recognizing the benefits of *Chữ Quốc ngữ*, Vietnamese intellectuals, who had a profound knowledge of Chinese and Nôm scripts, spearheaded the movement to promote the diffusion *Chữ Quốc ngữ*. They established private schools where students could learn French, Chinese, and *Chữ Quốc ngữ*. The Tonkin Free School [Trường Đông Kinh Nghĩa Thục], established in 1907, was an extremely popular school in Hanoi, with dozens of classes and hundreds of students. The schools taught Vietnamese at the primary level, as well as Chinese and French languages at secondary and high school levels.[24] In addition to establishing schools, some intellectuals used newspapers as a practical vehicle to diffuse *Chữ Quốc ngữ*. The *Đông Dương Tạp chí* [Indochine Magazine] (1913–1919) and *Nam Phong Tạp chí* [Nam Phong Magazine] (1917–1934) were outstanding magazines that had various activities designed to popularize *Chữ Quốc ngữ*.[25] Particularly, the *Association Pour la Diffusion du Quoc ngu* was established on May 25, 1938, for the purpose of teaching how to read and write *Chữ Quốc ngữ* as necessary skills for daily life.[26] The advantages of this broader subscription to *Chữ Quốc ngữ* included its contribution to reducing illiteracy among regular Vietnamese people and the victory of *Chữ Quốc ngữ* compared with the Chinese, Nôm, and French scripts, as it came to be the dominant script in Vietnam during the colonial period. Beyond the diffusion of *Chữ Quốc ngữ*, Vietnamese intellectuals encoded within the script itself elements of patriotism, making the script an emblem of independence and a significant factor in the movement for liberating the nation.

The Transliteration of Place Names in Vietnam During French Colonialism

Before the diffusion of *Chữ Quốc ngữ*, Vietnamese place names were written by nobles and the educated classes in Chinese and Nôm scripts, and were also spoken and transmitted orally among the regular people. When it comes to the traditional transmission of place name knowledge, historically, monarchies recorded the names of settlement areas, such as territories, districts, communes, villages, and hamlets. Under the reign of subsequent dynasties, the compiling of gazetteers was undertaken by scholars, and it is this genre of documentation that rises in popularity in the nineteenth century. In these works, the features of local geography, ethnicities, and boundaries were recorded for perhaps the first time.[27] Under the reign of the Nguyễn Dynasty (1802–1945), the place names of administrative units were recorded in detail across various works published by the Department of National History [*Quốc sử quán*].

Traditionally, Vietnamese place names were not christened after people. According to Ngô Đức Thọ, Chinese and Vietnamese people viewed proper names as sacred and believed that proper names have a direct relation with the destiny of people. Moreover, the worshipping of souls and spirits is another reason for not naming places after people because both cultures embrace the taboo of mentioning the names of deceased people. The conventions of taboos governing the use of royal people's names were initially established by Kings of Trần Dynasty in the thirteenth century.[28] According to this convention, existing names that had the same sound as the names of royal members had to be changed. The long continuity of Vietnam's history created a complicated system of place names that comprised official and colloquial names in which official names written in Chinese script expressed their dreams of prosperity, happiness, and success, while colloquial names described local geographical features.[29] This coexistence of spoken and written place names means that educated people could read and write place names written in Chinese while regular people could only speak place names.

Usually, place names comprise of two parts: the generic and specific. For instance, *Mount Everest* includes *Mount* (a generic part) and *Everest* (a specific part). The order of the generic and specific also changes depending on the language in which it is expressed, such that generics could appear before or after specifics to form compounds of X + generic or generic + X.[30] Vietnamese place names were written in the form of X + generic (e.g., Hà Nội Tỉnh, Ha Noi Province, 河内省) in documents written in Chinese script. However, when they are expressed in Vietnamese spoken language, this order is reversed as generic + X (e.g., Tỉnh Hà Nội, Province of Ha Noi). In the documents recorded in French, the generics were translated into French words, such as *thôn* (commune, 村), *xã* (hamlet, 社), *bản* (hamlet, 本), *tổng* (can-ton, 總), *huyện* (district, 縣), *phủ* (prefecture, 府), and *trấn* (territory, 鎮).[31] The coexistence of X+generic form (e.g., Grand *Lac*) and generic + *X* (e.g., *Huyen* de Tu Liem, *Rue* Balny) is effected by the multilingual situation in Vietnam in the colonial era in which three spoken languages and scripts consisting of Vietnamese, French, and Chinese coexisted. On colonial maps, the place names of hamlets and communes were written with their specific (e.g., [thôn] *Linh Đường*). But, in contrast, the names of higher administrative units were written with both their generic and specific parts (e.g., *Phủ Thanh Oai*).

Beyond the aforementioned political motivations, the diffusion, teaching, and learning of *Chữ Quốc ngữ* became an essential tool for recording Vietnamese place names in Romanized script. A Vietnamese syllable consists of a tone, an onset, a glide, a nucleus, and a coda. For example, the syllable *toán* [tuan5] (means Math or counting); it includes tone (') (*thanh sắc* in Vietnamese), the onset *t* [t], the glide *o* [u], the nucleus *a* [a], and the coda *n* [n]. While the tones and nucleus are usually present, the onsets, glides, and codas could be absent. The number of Vietnamese tones differs between dialects, with six tones in the North and four or five tones in the Central and South dialects.[32] These complexities must be considered in light of the inaccurately transliterated place names collected by European missionaries and explorers and which ended up on maps of Vietnam published before 1900. In particular, because of these differences between Vietnamese and Roman phonetic characteristics, the cartographers guided the rules of orthography on maps in ways that justified the elimination of tone marks. As we will see, this has important consequences.

Transliteration with Tones, Onsets, Glides, Nucleus, and Codas

Six tones in the Vietnamese language are divided into two groups. The first is the high-level group, which includes three tones: *ngang* in Vietnamese with no marker expressed, *sắc* marked by the mark (′), and *ngã* marked by the mark (~). The second group is the low-level group, which includes three tones: *huyền* marked by the mark (ˋ), *hỏi* marked by the mark (?), and *nặng* marked by the mark (.). Vietnamese accents "played a major role in the formation of words,"[33] so applying incorrect tones may change the meaning of words illustrated in the example of the syllable *ba* below.

ba [ba] means three
bà [ba2] means lady, grandmother
bã [ba3] means remainder, residue
bả [ba4] means bait
bá [ba5] means auntie
bạ [ba6] is an element of complex words, for instance, bậy bạ (bad)

Recording without marks to distinguish between the phonetic features of syllables is typical of place names displayed on the maps created before 1900, such as *Thanh -Tri* [Thanh Trì], *Phu—Ung—Hoa* [Phủ Ứng Hoà], *Huyên de Tư-Liêm* [Huyện Từ Liêm]. Otherwise, applying incorrect marks of tones can be traced to some place names (e.g., *Đàn Viên* [Đản Viên], *Thương Thanh* [Thượng Thanh]). As is typical of a monosyllabic language, each syllable in Vietnamese equates to a word with its meaning. Recording the inappropriate tones of place names therefore will change the semantics of names and, as a result, an inaccurate name could become a new name.

In terms of onsets, there are 22 initial sounds in Vietnamese syllables, which are represented by a single letter (e.g., /t/ is written as "t"), double letters (e.g., /c/ is written as "ch"), and triple letters (e.g., /ŋ/ is written as "ngh" when it appears before the following vowels: /i, e, ɛ, ie/). The presentation of place names on colonial maps shows the confusion experienced by cartographers in choosing letters to represent onsets. A single phoneme could be represented by two or more letters and vice versa. For instance, the aspirated consonant /t'/ was represented with the letters *th-* and *t-,* such as *Tạch Bích* [Thạch Bích] and *Thanh Oai.* In particular, the /k/ phoneme is expressed by the letter *k-* if it appears before the vowels and diphthongs /i, e, Ɛ, ie/, by the letter *q-* if it appears before the glide /w/, and in the rest cases, the /k/ phoneme is represented with the letter *c-*; for instance, the name of [Pagoda] *Trấn Quốc* [means Protect Nation] was recorded by *Trấn Cuốc* [means *Protect Hoe*].[34] The voiced stop /d/ phoneme was expressed with the letters *d-* in the name *Vạn—Diêm* [Vạn Điểm] and with the letter *th-* in the name of *Thường* [Đường] *Xuyên*. However, the letter *d-* represented the /d/ phoneme in the name *Dắt Sở* [Đắc Sở] and the /z/ phoneme in the name *Dịch Vĩ.* The letter *h-* represented the glottal stop /ʔ/ phoneme in the name *Phủ Thanh Hoai* [Phủ Thanh Oai], while the laryngeal /h/ phoneme is represented with the letter *h-* in the name *Hà Trì.*

The glide /w/ is represented with two letters, including with the letter *-o-* if it appears before the open and half-open vowels /a, Ɛ/ and with the letter *-u-* if it appears before the close or half-close vowels /e, ɤ̆, i, ie/, for instance, *huy, hoa, huân*, and after the /k/ phoneme in words *quê, qua, quân.*[35] On colonial maps, the glide was occasionally removed (e.g. *Hyên Kỳ* [Huyền Kỳ]). In terms of the nucleus, some diacritics are applied to identify the phonetic features of Vietnamese vowels, such as short and long, rounded and unrounded, and front and back vowels. In Vietnamese Romanized script, the distinction of vowels' features is expressed by letters as follows:

- The letter *ă* represents the */ă/* phoneme
- The letter *ê* represents the /e/ phoneme
- The letter *ư* represents the */ɯ/* phoneme
- The letter *ô* represents the */o/* phoneme
- The letter *ơ* represents the */ɤ/* phoneme
- The letter *â* represents the /ɤ̆/ phoneme

Exceptionally, the /i/ phoneme is represented with two letters: with the letter *y* if it appears in the syllabic with the /ʔ/ phoneme in the onset and with the letter *i* in the rest situations. On colonial maps, even if the /i/ phoneme appears after a glide, it still was recorded by the letter i (e.g., *Quình Lôi* [*Quỳnh Lôi*], *Huinh Cung* [Huỳnh Cung]). Not only did colonial-era cartographers make mistakes with the /i/ phoneme but also they did not distinguish between high front vowels and low back vowels (e.g., *Văn [Điền]* was recorded by *Vân* [*Điển*]; *Do/Gio [Lộ]* was recorded by *Dô/Giô [Lộ]*). In particular, three diphthongs /uo/, /ie/, and /ɯɤ/ are represented with various letters; for instance, the diphthongs /uo/ were written by *ua-* in the name *Thuyền Quang* and by *uơ-* in *Thuyền Quơn* and the diphthongs /ɯɤ/ were written by *uo-* in the name *Gia-Tuong* [Gia Thượng] and *Thuy-Chuong* [Thuỵ Chương].

In terms of codas, on colonial maps, the /ŋ/ was represented with three letters, including *-nh, -n,* and *-ng* in the names *Vĩnh Tuy, Vĩn [Vĩnh] Tuy, Xã Dàng [Xã Đàn], Liêng Đường [Liên Đường]*. The /k/ phoneme is expressed with two letters, including with the letter *-c* in the name *Cho-Moc* [Chợ Mọc -Mọc Market], by the letter *-t* in the name *Dắt Sở* [Đắc Sở]. Occasionally, the /k/ phoneme was removed (e.g., *Phu Lâm* [Phúc Lâm]). The /n/ phoneme, an open coda, was expressed by the letter *-m* in the name *Bạch Liêm* [Bạch Liên].

An analysis of place names written in Vietnamese Romanized script on colonial maps shows that colonial explorers made the maps themselves, at least initially. However, the missionaries sought the support of native people who had a profound knowledge of both Chinese and Nôm scripts, and they could speak French fluently. These distinguishing features of the Vietnamese colonial milieu result in various inaccurate place names on the maps made by Halais C. and Leclanger, for instance, *Guamla-Xa* [may be Quán La Xã], *Tsoun-cain* [Xuân Canh], *Gia-Quoit* [Gia Quất], *Yado-Xa* [Nghĩa Đô Xã], *Co-Giam* [Cổ Giám], *Linh-Quơn* [Linh Quang], *Xã-Dàng* [Xã Đàn].[36]

We can take the example of the Vietnamese Romanized place names displayed on the map, *An Nam Đại Quốc hoạ đồ (Tubula Geographica Imperri Anamitici)* (Taberd 1838b),[37] the map created by Bishop Jean-Louis Taberd (1794–1840) in 1838. The Vietnamese place names on his maps were written in isolated styles and the signs of tones were applied correctly, for instance, *Sơn nam trấn* [Territory of Sơn Nam], *Hải đông trấn* [Territory of Hải Đông], *Vịnh vạn ninh* [Vạn Ninh Bay]. Surprisingly, place names listed in the Taberd's dictionary were written in the same manner as Vietnamese orthography is today, such as *Bình Thuận [Bình Thuận Province], Qui Nhơn [Qui Nhơn Province], Hà Tiên [Hà Tiên Province]*.[38] Taberd mentioned that "in writing the native names in Romans characters," he followed "the method adopted alike by all missionaries of different nations for the last 200 years."[39] Moreover, he created maps based on the maps drawn "by engineers of the country gave it this direction. They know the country, they visit it every day and have measured all the windings of the river Laos."[40] However, the analysis of Taberd's work shows that he could not distinguish Vietnamese tones and syllables, for instance, *Tung-kin/Tong-kin/Thon-kin* [Đông Kinh], *Chơn lap/Chiêm lap* [Chân Lạp], *Quang tri* [Quảng Trị], *Quang duc* [Quảng Đức], *Dôngnai* [Đồng Nai], *Ke? chơ* [Kẻ Chợ], *Tông-ca?* [Sông Cả]. This observation demonstrates that Vietnamese people played an essential role in creating and revising Vietnamese Romanized place names and that colonial mappers themselves relied upon them.

In the maps created by colonists, the involvement of Vietnamese people in the revision of place names may be recognized clearly in the maps published after 1900 that rarely had any inaccurately written place names.[41] In his research into the activities of the Đông Kinh Nghĩa Thục [Tonkin Free School], Nguyễn Hiến Lê noticed that several staff members of the Tonkin Free School were employees of the Service Géographique de l'Indochine as well. They had made a "Vietnamese map on the white fabric . . . expressed accurately the names of rivers, mountains, citadels, resources, . . . by Chinese and Vietnamese scripts."[42]

In my opinion, this form of cartographic co-creation demonstrates the process of making colonial maps alongside the decolonizing of place names in the same period. But, also, the renaming of streets in Hanoi after the fall of French colonialism related to propaganda circulated in the newspapers about the reasons for (re)naming, which strove to erase the trauma of slavery and to remind the reader about their sense of patriotism, for which the foundation was the Vietnamese influence on place naming practices in colonial Vietnam.[43]

A Mosaic of Place Names

Besides transcribing Vietnamese names, the European colonists bore names based on the characteristics of geographical features, for instance, West Lake [Hồ Tây] is more significant than Hoàn Kiếm Lake [Return Sword Lake], so West Lake garnered the name *Grand Lac,* and Hoan Kiem Lake the name *Petit Lac.* Similarly, streets in the Old Quarter of Hanoi gathered certain commodities that are the roots of names given by the French, for instance, Rue des Voiles [Hàng Buồm], Rues des Vases [Hàng Chĩnh]. This situation created multilingual place names on colonial maps where names of administrative units were written in Vietnamese Romanized script, street names were recorded in French, and geographical names were represented in French, Vietnamese Romanized script, or comprised hybrid names. The mosaic of place names could also be written in Chinese, or appear in specific contexts, such as newspapers, administrative documents, novels, or other typical forms of publication. Because of various languages and scripts, a single place name could be written in many ways; for instance, Hanoi, the capital city of Vietnam today, was written in seven manners, including *Hà-Nội, Hanoi, 河内, Ha-noi, Hanoï, Ha-noï, and Hà Nội.*

Analyzing colonial cartography and related written material shows five main strategies for writing place names in the colonial era. First, place names were written with the Vietnamese characteristics that isolated syllables and applied signs for tones (e.g., *Bắc Ninh, Thị Cầu, Phủ Lạng Thương*). Second, place names were written without diacritics and signs of tones; otherwise, a hyphen was placed between two syllables of place names (e.g., *Thuong-tin* [Thường Tín], *Dong-van* [Đồng Văn], *Phu-ly* [Phủ Lý]). Third, the syllables of place names were written without isolation as well as without the diacritics of both the letters and signs of tones (e.g., *Hanoi* [Hà Nội], *Haiphong* [Hải Phòng], *Haiduong* [Hải Dương]). Fourth, French place names were transcribed according to the rules of Vietnamese phonetics; for example, the name *Square Paul Bert* was expressed by *Vườn hoa Bôn-Be.* Because the Vietnamese language lacks the /p/ phoneme, it was recorded by the /b/ phoneme represented by the letter *b*-. Moreover, there is no /l/ phoneme in the Vietnamese codas, so the name *Paul* was recorded by *Bôn*. And finally, place names were recorded with an implement component in Vietnamese and vice versa (e.g., *Hồ Tây,* Grand Lac).[44]

On a map, place names could be represented by transcribed forms of Vietnamese place names; for instance, *Tei Ho* is a name recorded based on Tây Hồ /tɤ̆i1 ho2/ [West Lake]. Otherwise, cartographers applied bilingual names, including one part being French and the other being the Vietnamese name. The linked word *ou* (meaning "or" in English) links two parts, for instance, *Grand Lac ou Tây-Hồ*, *Sông Nhĩ Hà* [Nhĩ Hà River] *ou Fleuve Rouge*. The choice in place names through which the reader might embrace one over the other demonstrates the cultural and historical complexity of these names. Names of administrative units were written in isolate-tonal Vietnamese as well as following Vietnamese grammar rules (e.g. *Sông Nhĩ Hà, Hồ Tây*) while street names were written in French, even though streets have long had Vietnamese names in the Old Quarter, for instance, *Rue Paul Bert, Rue des Tasses, Rue des Vases,* and so on.

This picture of place names in Hanoi and Vietnam more broadly was complicated by its multilingual society in which three languages and writing systems officially coexisted (Figure 2.1).

The headings for the information presented in the guidebook are written in Vietnamese and Chinese. The city guide recorded street names in three languages with different functions. Street names themselves were written in French (Boulevard Amiral Courbet and rue de l'Ancien Canal). The explanations indicating the start and the end of streets were written in Vietnamese. French names were

TÊN CÁC PHỐ 街名	BỀ DÀI 長度	DỊCH CHỮ TÀU 譯華文	TÊN TA
Amiral Courbet,(Boulevard) bắt đầu phố Fellonneau đến phố Paul-Bert.	775 m	姑啤街	Phố Quốc-Bê·
Ancien Canal (rue de l') bắt đầu từ phố Mới đến phố hàng Mã Mây.		舊渠街	

Figure 2.1 Street names in Hanoi in 1923 in a guidebook of the city (Nguyễn Bá Chính 1923, p. 27).

also translated into Chinese script. Interestingly, while the colonizers transcribed local place names, French names were transcribed into Vietnamese by words that have the sound nearest to the French name; for instance, *rue Courbet* was transcribed *Phố [rue] Quốc-Bê.*

The coexistence of several scripts caused various written versions of place names in Hanoi during the colonial era. This context is similar to place names in colonial Singapore, where the names changed depending on the ethnic and linguistic groups involved, resulting in multiple names for a place that varied from one group to the next.[45]

Conclusion

Under French colonial rule, Vietnamese place names were translated from Chinese and Nôm scripts into Vietnamese Romanized script and recognized in both spoken and written versions. In contrast to the process of erasing colonial street names that was executed immediately after the surrender of the French colonial government, the process of decolonizing place names and administrative names was completed during colonial times in Vietnam, determined by the writing place names in the colonial maps of Hanoi. Inaccurate place names were then revised following the rules of Vietnamese phonetics, and the revisers found the most appropriate Romanized characters to modify the sound but not necessarily the meaning in a suitable orthographic system. This differs from other colonial contexts. In colonial Singapore, although the Asian groups ignored the official place names christened by British colonists, the place names used by them were not accepted by the colonial government. In Vietnam's situation, the initial role of Vietnamese Romanized script as a proselytizing tool evolved to become a vehicle that could help colonists narrow down the role of Chinese script and to both penetrate and control Vietnamese society. However, Vietnamese patriots and intellectuals used Vietnamese Romanized script as a worthwhile vehicle for improving the education of people and to promote the patriotic and anticolonial movements. Vietnamese Romanized script emerged an effective tool in forging a pathway toward nationhood; it reminded citizens of their nationalism and built support for protesting domination and slavery by foreign invaders.[46] In other words, Vietnamese Romanized script "turned out to be a dangerous weapon" for decolonization.[47]

Beyond toponymy, decolonizing place names in Vietnam may be considered a competition between languages and scripts in which some and not others become politically and socially dominant. From this perspective, Vietnamese Romanized script helps to fortify the eventual victory of the anticolonial leaders and intellectuals who diffused Vietnamese traditional values to reach the peak of *đánh đuổi thực dân* [fighting and chasing out the colonists]. From the perspective of regular people, they benefited from learning Vietnamese Romanized script, therefore becoming more literate and educated.

Acknowledgments

This research has been done under the research project QG.21.55 "Evaluate the situation of naming streets in Hanoi city and suggest some plans of naming streets direct to a smart city" of Vietnam National University, Hanoi.

Notes

1 In the history of Vietnam, the term "colony" is used to refer to those who imposed colonialism on the land that is the territory of Vietnam today, including: (1) "Northern colonization [Bắc thuộc in Vietnamese]," which refers to the period of more than a thousand years when Vietnam was occupied by Chinese feudal dynasties from 111 BCE to 931 CE (see Trần Trọng Kim, Việt Nam sử lược [Brief History of Vietnam], Nhà xuất bản Tân Việt [Tân Việt Publisher] (1964), p. 44); and (2) "French colony [Pháp thuộc in Vietnamese]," which refers to the period when Vietnam was under French colonialism from 1883 to 1945. In this chapter, I am not referring to the Northern colonial period but focusing only on the period of French colonialism.

2 Robin A. Kearns and Lawrence D. Berg, "Proclaiming Place: Towards a Geography of Place Names Pronunciation," *Social & Cultural Geography* 3.3 (2002), pp. 283–302.

3 Brenda S. A. Yeoh, "Street Names in Colonial Singapore," *Geographical Review* 82.3 (1992), pp. 313–322.

4 O. Uluocha Nna, 2015, "Decolonizing Place-Names: Strategic Imperative for Preserving Indigenous Cartography in Post-Colonial Africa," *African Journal of History and Culture* 7.9 (2015), pp. 180–192, 181, 184.

5 Paul Monaghan, "'What Name?': The Recording of Indigenous Placenames in the Western Desert of South Australia," in *The Land Is a Map, Placenames of Indigenous Origin in Australia*, eds. L. Hercus, F. Hodges, and J. Simpson (Sydney: ANU Press, 2015), pp. 203–206.

6 Monaghan, "'What Name?'"; Nna, "Decolonizing Place-Names"; Petrus Angula Mbenzi, "Renaming of Places in Namibia in the Pre-Colonial, Colonial and Post-Colonial Era: Colonising and Decolonising Place Names," *Journal of Namibian Studies: History Politics Culture* 25 (2019), pp. 71–99.

7 Nna, "Decolonizing Place-Names," p. 185; Yeoh, "Street Names in Colonial Singapore," pp. 317, 320.

8 Yeoh, "Street Names in Colonial Singapore," p. 319. Also see Mbenzi, "Renaming of Places in Namibia;" Phùng Thị Thanh Lâm, "(Re)Naming Hanoi Streets after Historical People from Colonialism to Postcolonialism: Creation and Recalling Collective Memories," *Onoma* 51 (2016), pp. 25–44.

9 Nna, "Decolonizing Place-Names," pp. 181, 186.

10 Yun-Tsui Yeh, "'Erased Place Names' and Nation-Building: A Case Study of Singaporean Toponyms," n/d, p. 38. Also see Đinh Gia Thuyết, *Tiểu Sử Các Tên Phố Hà Nội* [Story of Hanoi' Street Names] (Hanoi: Hội Ái hữu Văn đoàn, 1951).

11 Reuben Rose-Redwood, "'Reclaim, Rename, Reoccupy': Decolonizing Place and the Reclaiming of PKOLS," *ACME: An International E-Journal for Critical Geographies* 15.1 (2016), pp. 187–206. Also see Kearns and Berg, "Proclaiming Place," p. 291; Maoz Azaryahu, "Street Names and Political Identity: The Case of East Berlin," *Journal of Contemporary History* 21.4 (1986), pp. 581–604; Maoz Azaryahu, "The Critical Turn and Beyond: The Case of Commemorative Street Naming," *ACME: An International E-Journal for Critical Geographies* (2011), pp. 28–33; Maoz Azaryahu, "Renaming the Past in Post-Nazi Germany: Insights into the Politics

of Street Naming in Mannheim and Potsdam," *Cultural Geographies* 19.3 (2012), pp. 385–400.

12 https://gallica.bnf.fr.

13 Central Population and Housing Census Steering Committee, The Viet Nam Population and Housing Census of 00:00 hours on 1 April 2019: Implementation organisation and preliminary results (2019), p. 32. https://www.gso.gov.vn/wp-content/uploads/2019/10/2.-ENG_Census-on-Housing-and-Population_2019_final.pdf

14 Michel Ferlus, "Langues et Peuples Viet-Muong," *Mon-Khmer Studies* 26 (1996), pp. 7–28; Nguyễn Tài Cẩn, "Thử phân kỳ lịch sử 12 thế kỷ của tiếng Việt [A Proposal of Periodization of Twelve Centuries of Vietnamese Language]," in *Một số chứng tích về ngôn ngữ, văn tự và văn hoá* [Some Evidence of Language, Writing and Culture] (Hanoi: Nhà xuất bản Đại học Quốc gia Hà Nội [Vietnam National University Press]), pp. 401–411, 401.

15 John DeFrancis, *Colonialism and Language Policy in Viet Nam* (The Hague: Mouton Publisher, 1977).

16 Nguyễn Tài Cẩn, "Điểm qua Vài Nét về Tình Hình Cấu Tạo Chữ Nôm [A Glimpse of the Situation of Nom Script Structure]," in *Một số chứng tích về ngôn ngữ, văn tự và văn hoá* [Some Evidence of Language, Writing and Culture] (Hanoi: Nhà xuất bản Đại học Quốc gia Hà Nội [Vietnam National University Press]), pp. 58–94, 58.

17 David Marr, *Vietnamese Tradition on Trial, 1920–1945* (Berkeley: University of California Press, 1984), p. 145.

18 The complication of languages and writing systems in Vietnam is illustrated chronologically by Chiung based on the work of DeFrancis. See Wi-wun T. Chiung, *Taiwan and Vietnam: Language, Literacy, and Nationalism* (Tainan: Center for Vietnamese Studies, 2020), p. 48; John DeFrancis, *Colonialism and Language Policy in Viet Nam* (The Hague: Mouton Publisher, 1977).

19 Nguyễn Văn Trung, *Chữ Văn Quốc Ngữ Thời Kỳ Đầu Pháp Thuộc* [Quốc Ngữ Literacy in Early French Colonialism] (Saigon: Nam Sơn xuất bản [Publisher Nam Sơn], 1975), p. 44.

20 Marr, *Vietnamese Tradition on Trial*, p. 146.

21 DeFrancis, *Colonialism and Language Policy*, p. 143; Marr, *Vietnamese Tradition on Trial*, p. 137.

22 The original text in Vietnamese is "bắt các công chức người Việt phải học quốc ngữ, nếu không, không được thăng ngạch trật." See Văn Trung, *Chữ Văn Quốc Ngữ Thời Kỳ Đầu Pháp Thuộc*, p. 15.

23 Trần Thị Phương Hoa, *Giáo Dục Pháp-Việt ở Bắc Kỳ (1884–1945)* [French-Vietnamese Education in Tonkin (1884–1945) (Hanoi: Khoa học Xã hội [Social Sciences Publisher], 2012), pp. 26, 37.

24 This original text in Vietnamese is "[trường] dạy cả Việt văn, Hán văn và Pháp văn. Ban Tiểu học chuyên dạy Việt văn, lên Trung học và Đại học mới dạy Hán văn và Pháp văn." See Nguyễn Hiến Lê, *Đông Kinh Nghĩa Thục: Phong trào Duy Tân đầu tiên ở Việt Nam* [Tonkin Free School: The First Duy Tan Movement in Vietnam], Saigon: Lá Bối, 1968), pp. 44, 51.

25 Đông Dương tạp chí [Indochine magazine] (1913–1919) was a weekly magazine published in Vietnamese romanized script and *Nam Phong Tạp chí* [Nam Phong Magazine] (1917–1934) was a monthly magazine published in both Vietnamese romanized and Chinese scripts. The magazines included translations of novels from western and Chinese languages into Vietnamese romanized script, administrative documents, lessons in French, history, mathematics, morality, and methods of writing by *chữ quốc ngữ*.

26 Trung tâm lưu trữ Quốc gia I [Centre of National Archives N.1], 2019. Truyền Bá Chữ Quốc Ngữ ở Bắc Kỳ [The Diffusion of Chữ Quốc Ngữ in North of Vietnam].

27 Lê Tắc, *An Nam Chí Lược* [Gazetteer of Annam] (Hue: Lao Động Publisher, Trung tâm Văn hoá Ngôn ngữ Đông Tây [Centre of East West Culture and Language], 2009); Nguyễn Trãi, *Nguyễn Trãi Toàn Tập—Dư Địa Chí* [Selected Works of Nguyễn Trãi—Gazetteer of Geography] (Nhà xuất bản Khoa học xã hội [Social Sciences Publisher], 1976).

28 Ngô Đức Thọ, *Nghiên Cứu Chữ Huý Việt Nam qua Các Triều Đại (Les Caractères Interdits Au Vietnam à Travers l'Histoire)* (Hanoi: Nhà xuất bản Văn hoá [Culture Publisher], 1997), pp. 19–20, 37.

29 Generally, inhabitants named places before the government officialized the names. This means that a place has two or more names, including colloquial ones and an official name. In Hanoi, for instance, the hamlet of *Nhân Mục* has the native name, *Làng Mọc* [Mọc Village]; the hamlet of *Yên Lãng* has the native name, *Kẻ Láng* [Láng Village]; the hamlet of *Thanh Lãm* has the native name, *Kẻ Thuần* [Thuần Village], etc. The meaning of specific parts in colloquial names is often "lost in the mist of time;" see Jan Tent, "Antecedent Generics: How Capes, Lakes, Mounts, and Points Are Named in the Antipodes," *Names A Journal of Onomastics* 64 (2016), pp. 148–157. On the colonial maps of Hanoi published before 1900, place names were written flexibly for colloquial and official names.

30 Tent, "Antecedent Generics," p. 148.

31 Bản is a term used to refer to internal administrative divisions in the mountainous areas of Vietnam, while the terms thôn or xóm [hamlets or villages] are often applied to divisions in the lower areas. A xã (commune) consists of one or more than one thôn." See Vũ Thị Minh Hương, Nguyễn Văn Nguyên, and Philippe Papin, *Địa danh và tài liệu lưu trữ về làng xã Bắc Kỳ* [*Répertoire des toponymes et des archives villageoises du Bắc-Kỳ*] (Saigon: Bibliothèque Vietnamienne, VI, École Française d'Extrême-Orient, 1999).

32 Hoàng Thị Châu, *Phương Ngữ Học Tiếng Việt* [Vietnamese Dialects] (Hanoi: Nhà xuất bản Đại học Quốc gia Hà Nội [Vietnam National University Publisher], 2009).

33 DeFrancis, *Colonialism and Language Policy*, p. 210.

34 Đoàn Thiện Thuật, *Ngữ Âm Tiếng Việt* [Vietnamese Phonetics] (Hanoi: Nhà xuất bản Đại học và Trung học chuyên nghiệp [Universities and Professional Colleges Press], 2007), p. 162.

35 Thiện Thuật, *Ngữ Âm Tiếng Việt*, pp. 178–179.

36 Charles Halais, Raphaël Enguéhard, and A. Simon, *Plan de Hanoï et de ses Environs* (Paris: Charles Bayle, 1891).

37 Jean-Louis Taberd, *Anam Đại Quốc Hoạ Đồ* (Calcutta: Oriental Lith. Press, 1838).

38 Jean-Louis Taberd, *Dictionaticum Latino-Anamiticum* (Serempore: J.C. Marshman, 1838), pp. i, ii, iii.

39 Jean-Louis Taberd, "Notes on the Geography of Cochin China," *The Journal of the Asiatic Society of Bengal* VI, Part II, No. 69 (1837), p. 740.

40 Jean-Louis Taberd, "Additional Notice on the Geography of Cochinchina," *The Journal of the Asiatic Society of Bengal* VII.76 (1838), pp. 317–324, 319.

41 Indochine française Service géographique, "Environs de Hanoï," *Gallica*, 1900. Retrieved November 10, 2022, https://gallica.bnf.fr/ark:/12148/btv1b531955539; Lécureur, "Plan de la ville de hanoï / levé sur le terrain et déssiné par le sergent Lecureur du Service géographique," *Gallica*, 1902. Retrieved November 24, 2022, https://gallica.bnf.fr/ark:/12148/btv1b10106810f; Service Géographique de l'Indochine, "Environs de Hanoï," *Gallica*, 1910. Retrieved November 26, 2022, https://gallica.bnf.fr/ark:/12148/btv1b53063741v; Service Géographique de l'Indochine, "Plan de la ville de Hanoï," *Gallica*, 1911. Retrieved December 4, 2022, https://gallica.bnf.fr/

ark:/12148/btv1b53063324v; Service Géographique de l'Indochine, "Plan de la ville de Hanoï," *Gallica*, 1936. Retrieved July 25, 2021, https://gallica.bnf.fr/ark:/12148/btv1b53066855g.

42 The original text in Vietnamese is: "một bản đồ Việt Nam bằng vải trắng, [. . .] ghi rõ tên sông, núi, châu thành, thổ sản . . . bằng Chữ Hán và chữ Việt." See Hiến Lê, *Đông Kinh Nghĩa Thục*, p. 55. A member of this school was in the service of the Service Géographique de l'Indochine, so it is not surprising the place names were edited and decolonized in the early of colonial-era. Unfortunately, I have not found any more details of this member as well as the maps edited in Chinese and Vietnamese by him.

43 Thanh Lâm, "(Re)Naming Hanoi Streets."

44 Nguyễn Bá Chính, *Hà Nội Chỉ Nam* [Guide de Hanoi] (Hanoi: Nghiêm Hàm ấn quán, 1923), pp. 95–137.

45 Yeoh, "Street Names in Colonial Singapore," p. 318.

46 Nguyễn Văn Trung, *Chữ Văn Quốc Ngữ Thời Kỳ Đầu Pháp Thuộc* [Quốc Ngữ Literacy in Early French Colonialism] (Saigon: Nam Sơn xuất bản [Publisher Nam Sơn], 1975), pp. 15–16.

47 DeFrancis, *Colonialism and Language Policy*, p. 258.

3 Ghanaian Surnames at the Crossroads

Empirical Insights

Osei Yaw Akoto and Maxwell Mpotsiah

Surnames play an important role in the Ghanaian naming system. They form an integral part of a child's identity and are often used to signify the family into which the child is born.[1] They also express intergenerational relationships with the child's ancestral lineage and further signify a person's identity and humanity. In this way, surnames are used to trace one's genealogy.[2] As a result, surnames form part of the unique onomastic signs that make up a person's socio-cultural existence.

In Ghanaian societies, a child acquires a name to legitimize his or her humanity and personhood.[3] Such names include a day name (which represents the day in which the child is born, circumstantial name (attributed to issues that surround the birth of the child), linearity name (the order in which the child appears at birth), and the family name.[4] The family name is usually obtained from someone dead or alive and who has earned respect from all members of the family. This is done to ensure that the child grows up to be like the person after whom they have been named.[5] The surname is an important aspect of the child's self-conceptualization and identity because it legitimizes the filial bond between the child and their father as expressed and recognized at formal and institutional levels.[6]

Surnames in Ghana are characterized by several transcultural onomastic features. The local surnames that are known to spiritually define a person are nonetheless being influenced by foreign naming practices. Mkhacani T. Chauke observes that such practices have become too common in Ghanaian society because "modern parents prefer to cultivate a spirit of onomastic individualism or creativity and search for special, perhaps unique name for their child" and, consequently, some parents choose foreign surnames "for aesthetic appeal or to satisfy a personal whim."[7] This development makes Ghanaian surnames characteristically multicultural and multilingual.[8]

There is a burgeoning scholarship on personal names across ethnic groups.[9] The studies have generally focused on naming practices, sources, classification, and the grammar of personal names. In his recent panorama on Akan personal names, Sekyi-Baidoo discusses the grammar of Akan personal names by

DOI: 10.4324/9781003414773-4

adopting ethnographic approaches.[10] Of the works on names in Ghana, surnames remain relatively understudied.[11] Interestingly, Sekyi-Baidoo devoted a chapter to surnames in his book; however, he used little data and focused on Akans alone. The present study uses a specialized large corpus of multiethnic surnames to investigate the structure and language origin of Ghanaian surnames. The findings of the study will serve as a useful cross-ethnic model on personal names, in general, and surnames in particular.

Language and the Ethnic Composition in Ghana: Conceptualizing the Onomastic Crossroads

Ghana is characterized by diverse languages from different linguistic spectrum of mixed genetic ancestry. Ghanaian society is multicultural and multilingual, with 80 different languages spoken nationwide.[12] Out of these languages, only 11 are officially written and studied in schools, although this figure can change at any time depending on the government's language-in-education policy and the pressure from language-in-education advocacy groups.[13] However, the other remaining local languages are socio-culturally productive as they serve various socio-political and economic purposes. The people of Ghana are also divided into ethnic groups. These ethnic groups are defined by traditions, history, ancestry, and culture. The broader purview of ethnic groups in Ghana is seen from the picturesque sense of an ethno-linguistic classification. This classification identifies five main ethnic groups in Ghana, and these are Akan, Ewe, Mole-Dagbane, Guan, and Ga-Adangbe. Each group comprises several units that reveal their genealogy. For example, Akan comprises Asante, Fante, Akwampi, Akyem, Akwamu, Bono, Ahafo, and so on. The Ewe groups also include Nkonya, Tafi, Logba, Sontrokofi, Lolobi, and Likpe. The Mole-Dagbane also involves the people of Mamprusi, Dagomba, Gonja, and Mossi. Ga-Adangbe encompasses Ga, Adangbe, Krobo, Shai, and others. Guan includes Awutu, Efutu, Nkonya, Anum, and Abiriw. These ethnic groups are found in different geographical locations in Ghana.

Ghana's onomastic trajectory spans the precolonial, colonial, and postcolonial eras. Each era is typified by unique onomastic peculiarities regarding name types, semantics, grammar, morphology, and pragmatics. In precolonial Ghana or before Ghana had contact with foreign cultures, Ghana was characterized by onomastic Indigeneity in that its onomasticon mainly contained resources from its cultural norms and beliefs. During the colonial era, Ghanaians were exposed to other name forms. This can be described as an onomastic or name contact because names from different cultures co-existed in the Ghanaian name ecology. The name ecology of Ghana during this era experienced changes through several adoptions and adaptations. Ghanaians with the new insights through exolinguistic names can foreignize their names morphologically, phonologically, orthographically, or semantically. In this period, some of the local names were deemed devilish or demonic.[14] These crossroads between Indigenous and non-Indigenous names can be seen in the dropping local names, adopting foreign names, foreignizing local names, creating their own distinct names in the colonial languages (this is possibly part of decolonizing their names, known as onomastic decoloniality), and by blending Indigenous and foreign names.

In postcolonial Ghana, there are new onomastic features that throughout the namescape. The discourse on decoloniality has affected the onomastic spheres of Ghana. People are now adopting local names, giving local names to their children, or creating their own names in foreign languages through diverse morphological processes such as clipping, compounding, blending, reduplication, among others. This decolonial wave has affected the onomastic practices of some religious institutions that historically would assign foreign names under the guise of religion to their followers after some religious rituals such as confirmation and baptism. This practice supports Aceto's report that

> members of many Roman Catholic orders choose new names to represent their new spiritual lives: Mother Teresa was born Agnes Gonxha Bojaxhiu. Black Muslims often drop their surnames, which they sometimes identify as "slave names" (typically Anglophone-derived names stemming from the history of European colonialism in the Atlantic region), when admitted into the Nation of Islam (Malcolm X [Shabazz] was previously named Malcolm Little; Muhammad Ali was born Cassius Clay).[15]

These observations suggest the complexity of the crossroads of the various naming systems of Ghana. Surnames now find themselves at the junction of foreignization, Indigenization, re-Indigenization (sankofa), and hybridity. Name givers, bearers, and users now have a significant task when it comes to surnames. For example, Muslims in Ghana are now naming their children without Islamic names after realizing that Arab names existed long before Islam. Christians, similarly, are reaching for Indigenous names.

The Ghanaian people have a unique name repertoire that reflects their cosmological or metaphysical worldviews. The various names of all kinds of entities generally define the namescape of a geopolitical environment. Ghanaian names are inextricably part of their ecosystem, social makeup, politics, and worldview. Names are considered repositories of the epistemologies, beliefs, and cultures of the people. This onomastic dilemma, then, is reflected in the practice of changing legal surnames in Ghana.

Corpus and Procedures

Names of Ghanaian students in a Ghanaian public university constituted the corpus for the study. The names of international students (as per their tags in the university information management system) were excluded from the data because the study focuses on Ghanaian students only. The names were downloaded from the university's database after seeking appropriate permissions. A Ghanaian surname in this study is the one borne by a student with a Ghanaian nationality tag or label as indicated in the database. The system allows one to filter to select only Ghanaian students and that was what was adopted in this case. The surnames were extracted from the main corpus of full names. The full name corpus was built in parallel, but for this study a specialized corpus of students' surnames, Surname Corpus-Ghana (SCG), was compiled. In all, there were 13,730 students' names identified. The SCG had 3,380 types and 1,595 tokens. AntConc was used to obtain quantitative profiles such as the tokens and types of the corpus. Each name that is distinct from all others is considered a type and an example of that name a token.[16] A token of a type can be a forename, middle name, or surname.

Both manual and computerized methods were adopted in identifying the language of origin of the surnames that were outside the knowledge of the analysts. There are several computer-based forms of software for surname analysis, including Namesor (https://namsor.app/), Ancestry ((https://www.ancestry.com/learn/facts), and Forebears (https://forebears.io/surnames). Forebears was mainly used to search for the sources of the surnames but the other platforms were employed for triangulation purposes.

In analyzing the structure of the surnames, we observed that some of them exhibited one or more themes. In dealing with the structure of the surnames, we regarded each element in the surname as a theme, such that a surname could have the following structure: Theme 1 + (Theme 2) + (Theme 3) + (Theme 4), where all the parenthetical themes are optional. In her paper, Kubayi used the terms First Surname' and Second Surname, but we depart from this practice because individual elements do not independently constitute surnames in complex or multiple surnames.[17] Of greater importance is the combination of the onomastic syntagms that constitute the full surname, as exampled by the full surname, IDDRISU NURUDEEN, whereas IDDRISU and NURUDEEN individually constitute only part of the full surname.

In the linguistic origin analysis, the source of each of the themes established the language combinatoriality. It was observed that some complex surnames were either homolinguistic or heterolinguistic. The terms homolinguistic and heterolinguistic refer to the language background of the surname type. Homolinguistic multiple surnames have the sources of the surname types in either Indigenous or foreign languages, while heterolinguistic multiple surnames comprise types from diverse languages (i.e., both Indigenous and foreign languages). The surnames ISSAH MOHAMMED and JEHU-APPIAH are examples of homolinguistic and heterolinguistic surnames, respectively, because in the former, all the types are from Arabic (which is a foreign language) but in the latter, JEHU is from a foreign language, while APPIAH is from an Indigenous language.

The study classified surnames as Indigenous or endolinguistic, and those from non-Ghanaian languages as foreign or exolinguistic. Three strategies were adopted in identifying the surnames in the datasets. First, we relied on our linguistic knowledge as members of the dominant ethnic group in Ghana to identify surnames from our respective ethnic groups. Interestingly, the first 25 most frequent surnames in the corpus were from Akan. Second, we engaged five members of the five dominant ethnic groups to help identify the surnames from their ethnic groups in the datasets. And third, we performed an electronic analysis that showed the sources of both Indigenous and foreign languages of the surnames.

A structural-origin analysis of a surname (e.g., KLEMESU-FRANKLIN) was then diagrammatically represented to enable our ability to analyze the data. The structural analysis then considers each onomastic constituent as a theme. This approach differs from the grammatical analysis of the name structure that classifies names into clauses, sentences, and phrases.[18] In cases such as GREGORIO DE-SOUZA and DE-FENZY BRUCE, they were counted as bithematic rather than trithematic names because "De" and allied ones were considered as prefixes, which is in line with the established practices in investigating surnames.[19] The structure, just like the origin of surnames, is not regulated within any legal or cultural framework in Ghana. Surname givers, changers, and adopters have that onomastic right to determine the thematic string of their surnames.

Analysis and Discussion

Structural Analysis of Surnames in Ghana

Surname typologies are determined based on several factors, including the language of origin, gender, ethnicity, meaning, etymology, and structure.[20] Adjah intimated that Ghanaian surnames are derived from family names, place names, ethnic names, and appellations.[21] This research thus offers a structure-based typology of surnames that appears to be unstudied in the literature on surnames and surnaming practices across the globe. The analysis of our corpus revealed insights into surnames in Ghana. It was discovered that in terms of structure, surnames can be classified into two major categories: monothematic and multithematic, containing one or more than one theme respectively.

Table 3.1 Structural forms of surnames.

Forms of Ghanaian surnames	*Freq./%*	*Examples*
Monothematic	11,783 (85.8%)	ZUBAIRU; ZANU; OSEI; YOROSEH; TWUM; XEXEMEKU; WUMPINI; OWUSU; WILSON; TWENTOH; AKOTO
Multithematic	1,947 (14.2%)	ADU ASAMOAH; BOYD BOATENG; OTOO KWEKU; GLAVEE ADZO, EPIS ELMIR

The entire corpus comprises 13,730 surnames, which correspond with the number of name bearers. Out of this number, 1,947 were structurally complex, such that they contained two or more themes. The multithematic names, therefore, constitute approximately 14% of the total corpus of surnames, and 11,783 or nearly 86% are monothematic surnames.

Monothematic Surnames

Monothematic surnames comprise one onomastic theme.[22] Traditionally, names are forms of address in libations, and all forms of traditional discourses that ensue between the dead and the living. As a key ethnolinguistic address or reference term, surnames in terms of their structure are required to be preserved.

Naming practices across ethnic groups in Ghana indicate that surnames are integral parts of personal names.[23] These surnames are either from patrilineal or matrilineal lineage, depending on the norms of the ethnic group into which a child is born. While there are Indigenous conventions and norms relating to the sources of surnames across some ethnic groups in Ghana, it is not clear whether there is a law that regulates the structure of surnames in any of the ethnic groups.[24] Ghanaian surnames are characteristically and traditionally monothematic, although there is evidence of bithematic surnames or prefixed surnames in some ethnic groups. It is, therefore, unsurprising that an overwhelming number of surnames (86%) are monothematic. The high density of monothematic surnames attests to the fact that statistical evidence shows that Ghanaian surnames are largely consistent with Indigenous naming structures.

Monothematic surnames are easy to identify, especially when their distinct multimodal features, such as positioning, uppercase, and commas, are provided. Given their simple structure, monothematic surnames do not pose challenges to people when they want to distinguish them from first and middle names. It appears that almost every Ghanaian has a kind of an intuitive onomastic knowledge that surnames are monothematic, so when addressing people the most common practice is to use monothematic elements. The onus usually lies on the name bearer to correct people by noting, for instance, that my surname is not MANU, but rather GYIMAH MANU.

Table 3.2 Examples of monothematic surnames.

Endolinguistic surnames	*Exolinguistic surnames*
DUODU; MIREKU; OKINE; OSAE; OWIREDU; TEYE; TUFFOUR; YAMOAH; ADADE; AFARI; AMPADU; ANNANG; ANNIN; ANSU	AJAGO; NEWMAN; SAMARI; PAYNE BARTIMEUS; ZAKARIA; BAVERE; GERALDO; ABUKARI; HAMMOND; AZINGA

Multithematic Surnames

Multithematic surnames contain two or more themes. The themes are entangled syntagmatic relations, such that they combine to reveal the ethnolinguistic identities of the bearers.

The corpus reveals the complexity and dynamics of Ghanaian surnames and that multithematic surnames, which divide into bithematic (14%), trithematic (0.17%), and quadrithematic (0.01%) surnames (Table 3.3).

Multithematic surnames share the same structural characteristics as full names in that every full name in Ghana contains at least two themes. This structural similarity makes it difficult to distinguish between surnames and full names because one's surname can be another person's full name. Some ethnic groups (e.g., Dagara) have socially defined surnames known as family names, automatic surnames, hereditary surnames, or fixed surnames, all being key components of the personal name structure. While one may not be restricted in choosing a first or middle name, the surname is predetermined as per social norms. However, there is no socially defined surname stock, meaning namers can exercise their onomastic rights to assign any structural form of surname to their newborns. Furthermore, mature people can also decide to choose any structural surname by swearing an affidavit for a change of surname.

Bithematic surnames, which are generally known as double-barrel or compound surnames, comprise two themes that are either hyphenated or non-hyphenated (Table 3.4).[25] Bithematic surnames involve two themes conjoined by either a hyphen or space. The analysis demonstrates that bithematic surnames are either hyphenated or non-hyphenated. While the two themes in the former are separated by a hyphen (e.g., ABAKAH-MENSAH; ABASS-ADAMS; ABAYIE-YEBOAH), those in the latter group are separated by a space (e.g., ABASS MUMUNI; ABDUL AZIZ; ABDUL HAADI). Hyphenated and non-hyphenated surnames in the corpus constitute 6.7% and 7.31% of the surnames surveyed. This shows that Ghanaians have a relatively greater preference for non-hyphenated surnames than hyphenated ones.

Table 3.3 Examples of multithematic surnames.

Structural types of multithematic surnames	*Freq./%*	*Examples*
Bithematic	1,924 (14%)	KISSEIH-DOGBE; KLEMESU-FRANKLIN; KOBIA-AMANFI; KOI-LARBI
Trithematic	22 (0.17%)	ANENYAH NTOSO ANDOH; OSEI-WUSU BONSU; MOHAMMED LILO SEIDU
Quadrithematic	1 (0.01%)	NANA A. SAM WOODE

Table 3.4 Hyphenated and non-hyphenated surnames.

Structural types of bithematic surnames	*Freq./%*	*Examples*
Hyphenated	920 (6.70%)	ABABIO-TROVEL; GYAMFI-BONSU; GYAMFI-KESSE; ABAKAH-GYAN; ABAKAH-MENSAH
Non-hyphenated	1,004 (7.31%)	AABABIO FOFIE; GREGORIO DE-SOUZA; ABASS MUMUNI; ABDUL AZIZ; ABDUL HAADI

Kubayi contends that bithematic names are global phenomena practised in countries such as Germany, Scotland, and Spain. She, however, noted that the phenomenon is increasingly getting widespread, especially since most women merge their maiden surnames with their husbands' surnames. This practice results in a surname policy or formula just like the case in Scotland, where it is a common practice for each person to have compound surnames. Kubayi reports that Scottish children are named based on a surname formula where the first and the second themes in the father's and mother's compound surnames are combined to constitute the surname of the newborn child.[26] For example, if a husband and wife with respective surnames GYASI BENYADE and HODZI AHMED give birth to a child, the child will inherit GYASI HODZI as his or her surname, being an example of a patro-matronymic surnaming system. Although there is no reported practice like this in Ghana, Husein and Alemna (1984) reported that the merging of spousal surnames is one of the causes of bithematic surnames in Ghana.

Another form of multithematic surname is the trithematic surname, which has three themes in a syntagmatic relation. Table 3.3 shows that trithematic surnames constituted 0.17% of the total corpus. The presence of trithematic surnames comprises a key onomastic consideration given that it has not been reported in the Ghanaian onomastic literature.[27] Trithematic surnames are derived from the two forms of bithematic surnames, namely hyphenated and non-hyphenated surnames. For example, in the trithematic surname OWUSU ADDAI-POKU, the theme OWUSU was added to the hyphenated name ADDAI-POKU to increase its complexity. Again, in ANENYAH NTOSO ANDOH, either ANENYAH was added to NTOSO ANDOH or ANDOH was added to ANENYAH NTOSO. We can therefore outline hyphen-embedded and non-hyphen-embedded trithematic surnames. The Indigenous names' structure in Ghanaian societies suggests that trithematic surnames are modern developments in their onomastic practices. This form may have arisen from the practice of merging surnames, as in the case of combining the surnames of couples, or a parent and a namesake where one of them probably had a compound surname. For example, an interaction with someone with a bithematic surname revealed that both the father's and the namesake's surnames were conferred on him. It is clear that Ghanaian surnames are now at a crossroad of complexity and this shift is already underway.

The last form of multithematic surnames is the quadrithematic surname, which comprises four themes. The only instance found in the corpus is NANA A. SAM WOODE. Although the figure is statistically insignificant, it is suggestive of the future of the surnaming in Ghana. Several factors increase complexity of the surnames of individuals, key among them marriage. Hence, as more people with complex surnames marry and merge their surnames, the more complex the structures become. This could be better managed through the adoption of an onomastic policy at the family, ethnic, or national level to control the structure of surnames as reported in Kubayi as is being practiced in Scotland. The absence of this policy accounts for multithematic surnames among the Dutch and German noble families where one bears the surname "Petronella Q.R. barones van den Bosch van Draken-teijn-Tutein van der Nahenius, gezegd Dratunus."[28]

Several factors account for multithematic surnames in general. Interactions with people who bear these names reveal diverse reasons: varying degrees of combining parents' surnames, and globalization of naming practices.[29] A key influential factor is onomastic prestige. Bemile on Dagara surnames asserts that they "can be used as a means of self-protection and of boosting of one's image, as a sign of prestige, [and] as a commemoration of the famous deeds of some-one."[30] There is also an instance when some people exercise their onomastic human right and legally change their surnames from monothematic to multithe-matic ones. Among some of the ethnic groups in Ghana like the Norwegians, "surnames are normally transferred to children from the family to which they belong. If a person at a later stage wishes to change his or her surname, he or she is relatively free to do so."[31] When these changes are made, some adopt complex surnames through various morphological processes like compound-ing and reduplication. People may change their surnames to reflect their new socio-professional status, to correspond with their future aspirations, to demon-strate their new religious identity, to manifest their transformed character, and to de-link themselves from their lineage.

Four possible factors are responsible for the complexity in surnames. This is a practice through which the surnames of both parents merge to preserve the patri-lineal and matrilineal ancestries. The second is the desire of people to add the appellations of their surnames to make the surname complex as in the surname SARFO-KANTANKA, where KANTANKA is the appellation of SARFO. The third is the practice of combining names from different sources as in birth order, circumstantial names.[32]

The multithematic surnames pose their own social and legal challenges. Names' users have difficulty determining what constitutes one's surname. The common format for surname structure is monothematic. People with multithe-matic surnames can easily be misrepresented in official documents. Several people have attested to this in processing passports and citing bibliographical information in scholarly works. For example, how can one know that NANA A. SAM WOODE is not a full name, but rather a surname? From the corpus of full names, the bearer of the surname NANA A. SAM WOODE has the full name as LYSANDA NANA A. SAM WOODE. Users of the name for any purpose would have difficulties determining the surname.

Studies have revealed onomastic multilingualism and multilingual onomastic ecologies that were largely attributed to religion, globalization, inter-marriages, prestige, education, among other factors.[33] Some surnames are linguistically homogenous while others are heterogeneous, culminating in Indigenous + Indigenous and Indigenous + foreign surnames, respectively. These indicate the evolving nature of Ghanaian surnames at the crossroads of multilingualism and multiculturalism. The language of origin analysis reveals two forms of surnames, namely, Indigenous/endolinguistic and foreign/exolinguistic surnames. The multilingualism and multiculturalism manifested in the surnames are critical to understanding onomastic change and evolution. Arko-Achemfuor notes that "every child or person is given a surname mostly from a prominent past or living member of the man's family."[34] It is important, therefore, for the origins of surnames to be investigated for a possible tracking system to establish when exolinguistic surnames entered the surname stock of Indigenous Ghanaian societies.

Endolinguistic Surnames

Ghana is linguistically and ethnically diverse. These are reflected in the surnames in the corpus where surnames from the diverse ethnic groups and languages were identified. Indigenous surnames convey multiple identities for their bearers. Davies maintains that "surnames are not only a signifier of family and, more so, kinship, but also of ethnicity, faith communities and regional identities, and that surname can serve to locate individuals in a wider web of international relationships."[35] Indigenous surnames socially construct a concentric identity, such that based on the surname, one can broadly be identified as an African in race or place of origin, which is the universal set that informs both national and ethnolinguistic identities. For example, it is arguably easy for the bearer of the surnames OWUSU and OSEI to be identified as Africans, and subsequently as Ghanaians, and then Akans in general, or Asantes in particular. Adjah offers deeper insight into the sub-identities within the ethnic group. She asserts that "most Ghanaian societies are made up of families, which form clans, and these clans make up ethnic groups. Children's names are based on the families into which they have been born."[36] Some of Indigenous surnames are conferred upon babies by parents exercising their ethno-legal rights and performing their parental responsibilities as well while others are later adopted through legal processes. Afful observes that "in recent times, however, some Ghanaians (usually educated to higher levels such as the university) are beginning to drop the 'institutionalized' names and maintain the 'house' name as an all-purpose name to demonstrate nationalism and cultural renaissance."[37] This practice can be described as onomastic sankofaism because it is a mark of re-connecting one's identities to one's ethno-ancestral roots.

The use of Indigenous surnames is an effort to preserve our true African and Ghanaian identities. Interestingly, people are dropping their western names for Indigenous ones, as attested to by Appiah: "not only are the adults with foreign names dropping them, they are refraining from giving their children foreign names."[38] This is inspired by decolonial thinking that challenges western imperialistic and hegemonic cultural practices that for years have made Africans demonize their own names and cultural practices. It is thus refreshing to note that the first 33 surnames in the corpus were Indigenous, with a total frequency of 4,833, constituting 35% of the corpus itself. The first exolinguistic surname in the corpus is MOHAMMED, which is ranked 34th with a frequency of 63. The status of MOHAMMED, which is religious in nature, reveals the enormity of Muslim influence on the surname practices in Ghana.[39]

In Ghana, Indigenous surnames are not religious signifiers or markers. Some names are typically associated with some of the world's major religions such as Christianity and Islam: ABRAHAM is associated with Christianity and its Arabic equivalent, IBRAHIM, is associated with Islam. Yet, it is common in Ghana to find Christians who bear Arabic names and for Christians, Muslims, or Traditionalists to bear Indigenous surnames.[40]

Exolinguistic Surnames

Exolinguistic surnames refer to surnames that originate linguistically from a non-Ghanaian language. They include foreign language-medium surnames that may have been formed by Ghanaians, such that they exist only in Ghana. When such surnames are in non-Ghanaian languages, they are considered exolinguistic. Generally, foreign surnames can be described as exolinguistic surnames (from non-Ghanaian languages) against endolinguistic surnames that are from Ghanaian languages. This further justifies why anglicized, Europeanized, foreignized, and westernized surnames are still considered Indigenous surnames.

Exolinguistic surnames have diverse linguistic origins, including Hebrew, French, Arabic, English, Portuguese, Dutch, German, Persian, Indian, Italian, Latin, Polish, Russian, and Greek. The foreign surnames in the corpus are linguistically diverse, reflecting the influence of exogenous cultures on Ghanaian surnaming practices. The language from which these surnames originated are avenues of multilingualism. Ghanaian languages in this case are recipient languages affected by other cultures' onomastic conventions, norms, and practices. Some exolinguistic surnames may have entered Ghana through expatriates, foreign movies, books, internet, among other sources. Usually, they undergo some changes orthographically/phonologically, especially because some languages are characterized by special characters or diacritics that are not used in any Ghanaian

Table 3.5 Examples of homolinguistic and heterolinguistic surnames

Foreign-Ghanaian surnames	*Foreign-Foreign surnames*
ROCKSON-AMPAH	FERGUSON LAING
OHIME-BUER	KAMIL MOHAMMED
AMPIAH-BONNEY	LACINA DIABY
LARRY-STONE,	ABDUL-NUMIN
GRIMMON-INTSIFUL	ABDUL KARIM

language. This suggests that some surnames undergo orthographic or phonetic adaptation to suit either English or Indigenous languages of the surname bearers. Generally, it is the former because the English orthographic and phonetic systems have significantly influenced the spelling of Ghanaian surnames across ethnic groups. It appears that foreign names have greater influence than the local ones in terms of prestige and numbers, so an attempt to trace any of such names from the region of higher concentration to the region of no or lower concentration can be described as a kind of onomastic colonialism.

The analysis revealed two forms of surnames that are homolinguistic and heterolinguistic, which respectively indicate onomastic monolingualism and multilingualism. The homolinguistic surnames comprise themes from Indigenous or foreign languages only, while the heterolinguistic ones were mainly from the multithematic names. In the former, we observe surnames from a foreign language combined or mixed with one from a local origin. Examples of these are presented in Table 3.5.

Heterolinguistic surnames further reveal the positioning of the languages involved in the surnames. Themes from Indigenous or foreign language groups are either placed at the initial or final stage. These positionalities have socio-onomastic implications, as they provide information on linguistic hierarchy through surnames. Why will one position an endolinguistic or exolinguistic theme at the initial, medial, or final stage of one's surname? The practice reveals information about ideologies and identities enacted through language choices for onomastic purposes.

The use of exolinguistic surnames has several repercussions for Africans in general and Ghanaians in particular. First, it overshadows our true Ghanaian identities, especially when that has to be determined based on names in the absence of the bearers. People will be quick to describe the bearers of the surnames BRUCE, CORNELIUS, JOHNSON, PETERS, and RESCUE as non-Ghanaians. Appiah argues that "science editors with knowledge of African names have a better chance of appropriately attributing, citing, and indexing the many articles that people with African names have written for English language journals."[41] Appiah argues that Africans bear Afro-linguistic names, forgetting that not only exolinguistic forenames or middle names abound but also surnames are borne by his African brothers and sisters. How will the editor, drawing on any of the surnames, be able to attribute a piece of scholarly work or invention to an African or Ghanaian?

Furthermore, the findings to some extent are threats to surname analysis. Abrahamse, Morrison, and Bolton indicate that

> surname analysis is a technique that demographers can use to identify the members of particular racial, ethnic, or language communities within a population. Insofar as a particular surname belongs uniquely to a particular racial or ethnic group, it is possible to identify its holder's probable membership in the group by using well-formulated surname dictionaries.[42]

Obeng observed that "names in Akan, as in other cultures, are pointers to their users' hopes, dreams, and aspirations; they may reflect their users' geographical environment as well as their fears, religious beliefs, and the philosophy of life and death."[43] If the trend continues, what will be the futures of Ghanaian names in general, and surnames in particular? As Appiah indicates,

> If you go to some African countries where Europeans settled, such as Ghana and Nigeria, you may find such surnames as Ferguson and Johnson. But Africans did not have such names before the Europeans arrived. They had their own naming system that reflected the numerous languages they spoke.[44]

The corpus shows that Ghana's surname system is not just at a crossroads but is going down an exolinguistic path. The presence of foreign names and specifically surnames is an approach or strategy to creating a new ancestry because people trace their genealogy along onomastic lineages or lines.[45] There are, therefore, ancestral surnames documented or undocumented but known to the people through indoctrination or the socialization process. The stool names in the chieftaincy institution in Ghana, for instance, serve as surname archives or museums to preserve them for generations who may not come to witness some of their society's endangered Indigenous surnames. Arguably, increasing foreignization is a threat to our Indigenous or endolinguistic surnames. Anderson adds that "surnames identify an individual not simply by her/his personal name but by a name, or part-name, that is the name of her/his family, a name with a genealogical source."[46] Using foreign surnames is a means to de-ethnocizing or redefining oneself.[47] It implies that the bearers of exolinguistic surnames detach themselves from their ethnolinguistic roots and attach themselves to the language of origin of the adopted surname because "one of the functions of surnames is to identify family or clan membership, asserting the child's association with the family and parentage."[48] Bearers of exolinguistic surnames conceal their ethno-onomastic identities to align themselves with the exogenous ones.

When and how did such surnames enter the traditional surnames repertoire in Ghana? There are several factors responsible for their presence in Ghana, including colonialism, religion, education, inter-marriage, economic motivation, and the onomastic prestige of foreign names (the achievements of past and present bearers of the names, identity construction, and onomastic ideology. Tuor asserts that "there is hardly any limit to the name-giver's motives, idiosyncrasy and invention or concoctionability."[49] The legal definition of Ghanaian may have also influenced the surname stock of the country. According to Ghana's 1992 constitution, among other things, one can become a Ghanaian by naturalizing, which has implications for the surname stock because the person may retain his original name while he obtains Ghanaian citizenship. Nyaaba's study of the surnames of Afro-Brazilians in Keta (Ghana) attests to this.[50]

It is worth noting that Ghanaians might have encountered some of the exolinguistic surnames through other languages. Onomastic scholars and philologists can investigate the roots of these surnames by appropriating the concept of contact tracing, where one may have examined the language (i.e., English) as the major source of surnames for Ghanaians, and the various languages that English might have come into direct or indirect contact with. The challenge with this approach is that it may not be easy for one to note the original forms of some English surnames because many English names find their original forms in other languages. These loan translations, such as Bon Coeur, became Bunker, and Wittenachts became Whitenecks.[51] In these cases, are we to consider the English equivalencies or the original names to determine the language of origin in the corpus?

Conclusion

The chapter mainly focuses on the thematic structure and the language of origin of the surnames in Ghana. The Personal Names Corpus was built from a specialized subcorpus of surnames for the present study. A computer-assisted analysis was adopted, and this was complemented by manual analysis. The analysis revealed the structural dynamics of Ghanaian surnames ranging from monothematic to quadrithematic. The study showed that monothematic surnames are predominant in the namescape in Ghana.

The language origin analysis also revealed the linguistically diverse surnamescape in Ghana. The surname inventory in Ghana is characterized by Indigeneity and foreignism, culminating in endolinguistic and exolinguistic surnames, respectively. Ghanaians have either directly or indirectly had contact with other cultures through both virtual and physical mediums. Factors that may have resulted in the influx of exolinguistic surnames in Ghana include interracial marriages, religion, and prestige. These practices have had both negative and positive implications. The negative implications include the extinction, endangerment, stigmatization of the Indigenous names, and the loss of our beliefs, all of which can inform surnames. On the other hand, the bearers of such names may sometimes be regarded as foreigners and hence accorded social, economic, and educational privileges. The findings open up a new wave in African onomastic research that calls for employing Indigenous and critical methodologies for critical onomastics. Ghanaians' choice and preference for foreign or exolinguistic surnames borders on identity politics and the hegemonic influence of foreign cultures.

The study has implications for the documentation of Ghanaian Indigenous surnames as well, and points to the need for a Ghanaian Surname Dictionary, which may recount the linguistic changes in Ghanaian surnames. The changes could be phonological, orthographic, semantic, grammatical, morphological, and pragmatic. The findings will be useful for computer-based tools (e.g., Namesor, Ancestry, Forebears, and Origin of Surnames) that catalogue Ghanaian surnames. It appears that almost all entries for Ghanaian surnames in these platforms are monothematic. The compilation was done on the assumption that surnames are generally monothematic. Cataloguers are likely to truncate multithematic surnames, which will eventually result in misrepresentation of surname-based identities.

The present study does not provide an ethnographic account from surname givers, surname bearers, and individual and institutional surname users about their attitudes toward structural and language-based surnames. It will, therefore, be insightful for further studies to interview Ghanaians to ascertain their motivations for giving these surnames, attitudes toward these surnames, and challenges in using them. It is also recommended that a comparative study be undertaken on Ghanaian surnames across the precolonial, colonial, and postcolonial eras to establish the evolution that has occurred over time and across cultures and religious identity.

Notes

1 Olive Akpebu Adjah, "What Is in a Name? Ghanaian Personal Names as Information Sources," *African Research and Documentation* 117 (2011), pp. 3–17; Sikheto Joe Kubayi, "On Double-Barrel Surnames: Perceptions from Young Unmarried Women in Rural Limpopo Province, South Africa," *Nomina Africana: Journal of African Onomastics* 36.1 (2022), pp. 33–45.

2 Kofi Agyekum, "The Sociolinguistic of Akan Personal Names," *Nordic Journal of African Studies* 15.2 (2006), pp. 206–235.

3 Yaw Sekyi-Baidoo, *Akan Personal Names* (Accra: University of Ghana Press, 2019).

4 Agyekum, "The Sociolinguistic of Akan Personal Names;" Abraham Kwesi, "A Study of Personal Names among the Bikpakpaam (the Konkomba) of Ghana: The Linguistics, Typology and Paradigm Shifts," *Language Sciences* 66 (2018), pp. 15–27; Tolitey Emmanuel Cheditey, "A Socio-Cultural Analysis of Bɪrfor Personal Names," PhD dissertation, University of Education, Winneba, 2019; M. E. Kropp Dakubu, "Personal Names of the Dagomba," *Research Review* 16.2 (2000), pp. 53–65.

5 Kwesi, "A Study of Personal Names among the Bikpakpaam," pp. 15–27; Sekyi-Baidoo, *Akan Personal Names*.

6 Sekyi-Baidoo, *Akan Personal Names.*
7 Mkhacani T. Chauke, "Personal Names and Naming Practices among the Vatsonga," *The Anthropologist* 19.1 (2015), pp. 303–312, 306.
8 Sekyi-Baidoo, *Akan Personal Names.*
9 Bisilki, "A Study of Personal Names among the Bikpakpaam," pp. 15–27; Agyekum, "The Sociolinguistic of Akan Personal Names;" Sekyi-Baidoo, *Akan Personal Names*; Cheditey, "A Socio-Cultural Analysis of Bɪrfor Personal Names;" Dakubu, "Personal Names of the Dagomba," pp. 53–65.
10 Sekyi-Baidoo, *Akan Personal Names.*
11 Adjah, "What Is in a Name?" pp. 3–17.
12 Mary E. Kropp Dakubu, *The Languages of Ghana* (New York: Routledge, 2015); Felix K. Ameka and M. E. Kropp Dakubu, eds., *Aspect and Modality in Kwa Languages* (Amsterdam: John Benjamins Publishing, 2008).
13 Agyekum, "The Sociolinguistic of Akan Personal Names."
14 Sekyi-Baidoo, *Akan Personal Names.*
15 Michael Aceto, "Ethnic Personal Names and Multiple Identities in Anglophone Caribbean Speech Communities in Latin America," *Language in Society* 31.4 (2002), pp. 577–608, 584.
16 D. Kenneth Tucker, "Distribution of Forenames, Surnames, and Forename–Surname Pairs in the United States," *Names* 49.2 (2001), pp. 69–96.
17 John M. Anderson, *The Grammar of Names* (Oxford: Oxford University Press, 2007); Kubayi, "On Double-Barrel Surnames," pp. 33–45.
18 Bisilki, "A Study of Personal Names among the Bikpakpaam," pp. 15–27.
19 Anderson, *The Grammar of Names*; Sekyi-Baidoo, *Akan Personal Names.*
20 Anderson, *The Grammar of Names*; Adjah, "What Is in a Name?" pp. 3–17. Sekyi-Baidoo, *Akan Personal Names*; Ivana Fijok, "Onomastics as Evidence of Linguistic Influence," PhD dissertation, Josip Juraj Strossmayer University of Osijek, 2012; Ivana Fijok, "Onomastics as Evidence of Linguistic Influence," PhD dissertation, Josip Juraj Strossmayer University of Osijek, 2012.
21 Adjah, "What Is in a Name?" pp. 3–17.
22 Anderson, *The Grammar of Names.*
23 Adjah, "What Is in a Name?" pp. 3–17; Sekyi-Baidoo, *Akan Personal Names.*
24 Bisilki, "A Study of Personal Names among the Bikpakpaam," pp. 15–27.
25 Donald Ngmeneabangne Tuor, "Influences and Implications of Anthroponyms: The Case of Suuryee among the Dagara of Northern Ghana." PhD dissertation, Ghana Institute of Journalism, 2015; Kubayi, "On Double-Barrel Surnames," pp. 33–45.
26 Kubayi, "On Double-Barrel Surnames," pp. 33–45.
27 Kwasi Ansu-Kyeremeh, "Communicating Nominatim: Some Social Aspects of Bono Personal Names," *Research Review* 16.2 (2000), pp. 19–33; Akwasi Arko-Achemfuor, "Naming of Children and Meaning of Names among the Akan of Ghana: Defining Identities?" *Southern African Journal for Folklore Studies* 28.1 (2018), pp. 1–14; Bisilki, "A Study of Personal Names among the Bikpakpaam," pp. 15–27; Agyekum, "The Sociolinguistic of Akan Personal Names," pp. 206–235; Sekyi-Baidoo, *Akan Personal Names.*
28 Kubayi, "On Double-Barrel Surnames," pp. 36.
29 Aceto, "Ethnic Personal Names," pp. 577–608; Adrian Koopman, "Surname Dynamics in Avian Nomenclature," *Nomina Africana: Journal of African Onomastics* 31.2 (2017), pp. 141–152; Kubayi, "On Double-Barrel Surnames," pp. 33–45.
30 Cited in Donald Ngmeneabangne Tuor, "Influences and Implications of Anthroponyms."

31 Helleland, Botolv. "The Great Onomastic Divide in Norway: The Standardizing Problem of Settlement (Farm) Names after 400 Years of Danish Influence." In *Names and Their Environment: Proceedings of the 25th International Congress of Onomastic Sciences*, Glasgow, 25–29 August 2014, Vol. 4, Theory and Methodology. Socio-onomastics, edited by Carole Hough and Daria Izdebska. Glasgow: University of Glasgow, 2016.
32 Bisilki, "A Study of Personal Names," pp. 15–27; Sekyi-Baidoo, *Akan Personal Names*; Agyekum, "The Sociolinguistic of Akan Personal Names."
33 Veronika Štěpánová, "How Should These Names Be Pronounced? Specific Phonetic Features of Proper Names in Czech," in *Names and Their Environment*, eds. Carole Hough and Daria Izdebska (Glasgow: University of Glasgow Press, 2014).
34 Arko-Achemfuor, "Naming of Children," p. 3.
35 Hayley Davies, "Sharing Surnames: Children, Family and Kinship," *Sociology* 45.4 (2011), pp. 554–569, 556.
36 Adjah, "What Is in a Name?" p. 5.
37 Joseph Benjamin Archibald Afful, "Address Forms among University Students in Ghana: A Case of Gendered Identities?" *Journal of Multilingual and Multicultural Development* 31.5 (2010), pp. 443–456, 448.
38 Bernard Appiah, "African Names: A Guide for Editors," *Science Editor* 33.1 (2010), pp. 15–17, 16.
39 Appiah, "African Names," pp. 15–17.
40 Bisilki, "A Study of Personal Names," pp. 15–27.
41 Appiah, "African Names," p. 15.
42 Allan F. Abrahamse, Peter A. Morrison, and Nancy Minter Bolton, "Surname Analysis for Estimating Local Concentration of Hispanics and Asians," *Population Research and Policy Review* 13.4 (1994), pp. 383–394, 383–384.
43 Samuel Gyasi Obeng, "Akan Death-Prevention Names: A Pragmatic and Structural Analysis," *Names* 46.3 (1998), pp. 163–187, 163.
44 Appiah, "African Names," p. 15.
45 Kubayi, "On Double-Barrel Surnames," pp. 33–45.
46 Anderson, *The Grammar of Names*, p. 97.
47 Koopman, "Surname Dynamics in Avian Nomenclature," p. 147.
48 Kubayi, "On Double-Barrel Surnames," p. 34.
49 Tuor, "Influences and Implications of Anthroponyms," p. 20.
50 Ali Yakubu Nyaaba, "Reverse Diaspora and Reverse Acculturation: A Re-examination of Afro-Brazilian Surnames in Keta (Ghana)," Paper presented at the Africana Studies Interdisciplinary Conference, James Madison University, Virginia, USA. October, 18–19, 2012.
51 Koopman, "Surname Dynamics," p. 147.

4 Policy, Education, and Storytelling

Approaches to Decolonizing Canadian Toponymy

Lauren Beck

Given the complexity of socio-political and public discourse at the intersection of identity across the Americas, decolonizing place names presents several fundamental challenges that range from the permanent-seeming quality of names to the government and institutional processes for implementing them.[1] The toponymic status quo reflects both centuries of colonization and the people who made themselves naming authorities. As a result, the landscape remains whitened and masculinized through commemorative names that celebrate the values and worldviews of white men.

In the last few years, scholars have concluded that place naming practices throughout the western world are outdated, perpetuate colonialism, and exclude most of the people who live in a country such as Canada.[2] A preference for commemorative naming relating to historical figures, war dead and veterans, and political figures continues to impact naming from a gendered perspective.[3] Increasing awareness that certain place names can and should change has swelled public knowledge about the subject in recent years and is empowering communities across Canada to take action. Public consciousness about place names as extensions of our identities and lived experiences has also increased support for finding ways to better include underrepresented groups in our names, and we can look to other settler-colonial and postcolonial milieux, for instance in Africa and Latin America, for approaches toward this objective.[4]

In this chapter, I examine what decolonizing place naming practices implies before pursuing three key approaches toward decolonization: policy, education, and storytelling. The case study of Canada will serve as a model for other settler-colonial states that suffer from similar demons.

Decoloniality and Place Name Science

In Canada, colonial names date from the European invasion in the late 1400s when names that celebrated colonial monarchies and Catholic saints began to occupy European maps (i.e., Cape Breton, Nova Scotia). They were later augmented with names commemorating colonial officials and politicians (i.e., Vancouver, British

DOI: 10.4324/9781003414773-5

Columbia), most of which are rendered in English or French. Other names may relate in some way to Indigenous words but have been Anglicized or Frenchified (i.e., Québec), and themselves may not necessarily comprise Indigenous toponymy. In the twenty-first century, as the country grapples with its history of violence toward Indigenous peoples, in addition to centuries of sexism and racism toward women and people of color, clearly Canadians' values have shifted away from a desire to commemorate the perpetrators of colonial violence.

Decoloniality has more than one definition and application when it comes to places.[5] In many contexts, decolonization will be measured by the removal of colonial names. For others, it will be measured by the restoration of precolonial Indigenous names.[6] In other cases, it will be accomplished when names better reflect constituencies beyond white men.[7] Decolonization is also possible without Indigenization (i.e., replacing a colonial name with a descriptive one, such as Red River for the red color of the rocks), and Indigenization can occur without decolonization, for example, by embracing Indigenous names for First Nations as per the United Nations Declaration on the Rights of Indigenous People (UNDRIP) and the Truth and Reconciliation Commission's (TRC) 2015 Calls to Action.[8] As these different approaches demonstrate, when the will exists to change a name, a multitude of decolonial solutions can be marshaled.

In the western world, attempts to decolonize names have focused on reinstating Indigenous names and removing racist or colonial ones now deemed problematic, albeit with limited results.[9] Little progress has been made in dealing with sexist names, the inequalities created by the preponderance of names referring to white men, restoring precolonial names, or crafting names that reflect other populations impacted by settler-colonial naming practices (i.e., Black and Afrodescendent groups affected by slavery).[10] This problem was flagged 70 years ago by Watson Kirkconnell, who drew attention to the ways that names underlined cultural stratification in Canada, but little action has been taken.[11] Progress seems to be more possible at the local level, as demonstrated by intercultural toponymy projects such as the Taku River Tlingit First Nation's atlas of its place names (https://trt.geolive.ca/), and the nimbleness with which municipalities can adjust their naming practices. Progress is rarer at the national level in terms of decolonizing a country's name—Aotearoa (New Zealand) and Turtle Island (Canada/North America) being nascent examples that have had varying degrees of success in terms of replacing a colonial name.[12]

Indigenous people in particular have made significant strides in decolonizing names, whether through activism to encourage embracing an Indigenous name for a First Nation and its reserve, in the case of Elsipogtog First Nation, where the Indigenous majority rejected a colonial name and embraced its precolonial one, or which gestures toward precolonial times, as in 1492 Land Back Lane in Ontario.[13] Other attempts to see colonial names such as the Saint John River discarded in favor of precolonial ones have met resistance from different levels of government, which highlights the need to find means beyond government policy and regulations to change names.[14] Changing place names is part of a larger process of tribes and First Nations reclaiming and asserting their powers of self-determination and collaborative decision-making. Their political and cultural sovereignty differentiates Indigenous nations from other underrepresented groups also impacted by colonial naming practices.[15]

Importantly, the membership of place name boards such as the Geographical Names Board of Canada (GNBC) has been and continues to be predominately male and white in terms of their composition. This imbalance means that decision makers who influence naming may not be attuned to the inequalities provoked or sustained by their naming decisions and the procedures that inform them. In contrast, the Indigenous Place Keeping committee of Canadian Municipalities is primarily comprised of Indigenous people and women. At present, place name decision-making usually occurs through local, regional, and national processes using a place naming authority or committee invested with the authority to name. In Canada, municipalities determine how their infrastructure, streets, and parks are named, whereas the provinces and territories oversee the denomination of geographical features and communities. They report their activities to the GNBC, which sets best practices related to those of the United Nations Group of Experts on Geographical Names (UNGEGN), with which the country is affiliated.

The unquestioned structure of naming authority in Canada will inevitably reinforce the settler-colonial status quo. An important approach to decolonizing names will be building capacity within and outside of government to influence naming, particularly by empowering citizens to determine and vocalize names. The development of tools that any community can use to influence place naming will support decolonial objectives, and thus circumvent common barriers facing citizens (i.e., unclear name change processes, lack of policy and resources for researching names, etc.).[16] Like many Indigenous communities and activists, municipalities in places such as Canada have expressed tremendous interest in decolonizing names, but they struggle to make progress, and at other levels of government there also exists a lack of leadership, resources, tools for data, expertise, and policy to effectuate change.[17]

Through the following three instruments for decolonizing names, I demonstrate how they work and can make a meaningful contribution to decolonization through several examples of name changes and related processes.

Policy (Re-)Development

Until recently, no location exists where naming policies can be researched or consulted for the benefit of municipalities seeking to revise their policies by seeing what other communities have accomplished. Digitized datasets for both names and naming policy remain underdeveloped and limited by systemic factors such as the inability to use non-Latin fonts and the siloed nature of government websites at all levels. To address this issue, in 2023 my research team created a searchable database of municipal policy from across Canada's 4,500 incorporated settlements.[18] With this data, we concluded that only 4% of Canadian municipalities have a place name policy, which points to broader need for policy development when the matter of dealing with problematic names is facing scrutiny in the twenty-first century. Citizens often lack accessible tools to research place names in their communities, just as municipal staff typically lack starting points for understanding the state of their community's names and commemorative material. Before offering a solution for this problem, I provide an overview of the policy trends at this level of government.

The State of Municipal Naming Policy in Canada

Our analysis exposed broader problems within existing policy. First, 15% of policies commit to addressing colonial names, whether through dealing with a street such as Colonization Road, commemorations that are no longer desirable, in the case of John A. Macdonald, or making sure that no more colonial names are awarded. Second, 17% of policies commit to diversifying the gendered nature of their names, and this for the most part is a future-looking policy aimed at new named places and streets. Third, 18% of policies commit to addressing racism within their names, which necessarily involves changing racist names and requiring new ones to commemorate, for instance, Black presence and contributions. Fourth, 6% of policies commit to consulting marginalized communities about existing or future names, which is to say, approaching an adjacent First Nation or a women's activism group to consult about existing and future names. And fifth, 42% of policies identify "historical figures" as a naming priority.[19]

This last priority exemplifies common policy language that can be detected at other levels of government. For instance, many policies embraced by settlements reflect the GNBC's *Guiding Principles* for commemorative naming, both in Principle 5 and Appendix 3 on Commemorative Naming. The influence that the GNBC has in this respect is demonstrated through, for instance, the recommendation that names of this nature have a connection between the person and the locale being named, which we observed to be a standard policy point in Canada's settlements. And Principle 6 advises that the names of "pioneers," war dead, and names relating to historical events in the area be prioritized. Anecdotally, municipal staff have confirmed in exchanges with me that their policies have been inspired by those of other levels of government, and, from a different perspective, national naming authorities in Canada have admitted to me that their guidelines have long grown outdated.

Our high-level analysis of municipal naming policy in Canada revealed two significant cleavages: first, municipalities need policies that guide naming in the direction of decoloniality and inclusion, and second, most existing policies remain dated and need revision. The following best practices from existing policies were also notable and offer a meaningful pathway toward decoloniality:

Connect Policy to Other Value Statements

Most if not all municipalities have policies that guide expectations for behavior and decision-making practices when it comes to gender, race, and disability. These are designed to impact, for example, new infrastructure projects so to ensure that people do not face barriers to entering and using them. Communities are also legally bound by powerful documents that relate our identities to values and principles, such as the Canadian Charter of Rights and Freedoms. Some municipalities have made group-specific pledges—for instance, adopting the UNDRIP and responding to the TRC's Calls for Action, both of which specify the importance of using Indigenous-embraced and not settler names for spaces relating to or occupied by Indigenous people. When a settlement has already articulated priorities for inclusion and fighting coloniality through these means and has implemented them in areas under its purview unassociated with naming, it is logical to do the same when it comes to naming spaces. Connecting those policies to a municipality's naming policy will help offset and even treat the colonial condition of a settlement's names by fomenting greater inclusion. Naming decisions will then be informed by these commitments to values.

Address Demographic Gaps Deliberately

Most if not all communities in Canada have an underrepresentation of women and people of color, as well as Indigenous people, in municipal names. This is due, in part, to commemorative naming and the demographic that has traditionally done the naming. Most municipalities lack the resources to perform a detailed analysis of the demographics commemorated with names, yet it should not be difficult to ascertain where gaps are likely to exist when communities study their existing demographic composition. Commemorative names are an important way of connecting places to cultural, ethnic, religious, and linguistic groups within the community. Through named spaces, people feel seen and can see someone like themselves celebrated. By identifying which groups to promote from a community's existing population, policy can set naming priorities that specifically address those gaps, for instance specifying that women will be prioritized for all future street names.

Good policies articulate how changes may be determined and implemented. In the case of colonial names flagged as problematic by the community, settlements must develop procedures through which those names will be changed or recontextualized so that names fall in line with the municipality's stated values. In tandem, the process for naming usually requires municipal staff to work with its council to finalize naming decisions. An important ingredient for policy process is to ensure that broader demographics influence naming, whether through membership on the municipality's naming committee, public consultation, and regular requests for name proposals, or through demographically targeted, compensated forms of consultation.

Another perspective on addressing gaps is through language. The GNBC advises that names should be easy to pronounce and spell, as well as rendered in English or French. This requirement has broadly impacted policy at the municipal level in exclusionary ways that ignore how English or French words can be difficult to pronounce and spell for non-native speakers of those languages. The requirement that Romanized text be used, which frustrates names that require accents or diacritics, is another barrier to addressing demographic gaps. In this sense, good policy should state that names will not be excluded due to the perception that they may be difficult to pronounce or spell. Doing so will overcome presumptions about which names white demographics will find difficult to articulate, which fundamentally could undermine including names that refer to non-white demographics or those with varying levels of competency in Canada's official languages.

Broaden Categories for Commemorative Names

When policies prioritize or define commemorative names as a category, they usually list typologies such as "historical figures" that favor white men. To combat the status quo, municipalities should consider eliminating the requirement that commemorative names have to be historical in some way and, in the balance, we should be uneasy with linking identity to "history" in a settler-colonial context because doing so in places such as Canada will only reinforce the gender and race gaps. Ideally, names should be inspired by an individual's artistic, cultural, community, economic, educational, literary, political, military, religious, scientific, social, or athletic achievements.

"Firsts" are a common justification for commemorating a person, so a community might consider awarding names relating to the first business operated by an Indigenous person, the first teacher of color in the community, the first woman mayor, and so on. Specifying these categories will allow the proposers of new names to have broader sources of inspiration, for instance, by explicitly including categories such as the cultural industries or athletic achievements. Communities might also decide to move away from commemorative naming altogether so to prioritize names that refer to the landscape, ecosystem, industry, and so on.

Prioritize Marginalized Groups

Policies can designate certain groups that the community wishes to see better represented by stating that some new or revised names will relate to designated groups or, in contrast, that new names will not normally celebrate white men due to the preponderance of existing ones that represent this demographic. Making explicit how marginalized groups will be better accounted for in a community's names is an important area that a good policy should address. In parallel, existing names that refer to marginalized people should be reviewed to ensure that they reflect the policy's values statement. There are many resources that may support this action beyond education, explored in due course, including working with cultural agencies and institutions, such as temples and mosques, to identify problems with the community's existing named spaces, and to devise protocols for how marginalized groups will be prioritized. This step is particularly notable when neighborhoods become occupied by demographics who desire to live close to where they worship or go to school. In this sense, communities need to have a plan in place to find ways to use names to reflect the populations who live in that area of the community and which may not reflect the original demographic who lived there when the community was first planned—as exampled by Little Britain, which subsequently became Little Italy after Italian immigrants concentrated in that neighborhood in the mid-twentieth century, in Toronto (Ontario).

Use First and Last Names

As a study of Sackville, explored in due course, recently demonstrated, 18% of its 133 street names refer to women, yet none used their surnames.[20] This tendency foreshadows a significant barrier to studying place names from a gendered perspective because women, historically, have changed their surnames upon marriage. With their surnames removed from street signs, citizens often have no idea to whom the name refers, and many make assumptions that connect names such as Charlotte to noble women, that is, Queen Charlotte, when in the case of Sackville this street refers to the daughter of an eighteenth-century settler and landowner. In contrast, all names that refer to men in this community used their surnames and sometimes their first name as well. One important step that communities can take, in concert with educational institutions, is to research names that relate to women and revise street names so that they reflect their full names. There are many ways this could be accomplished—aside from an official name change, which would require updates spanning from driver's licenses to postal authorities—for instance, unofficial names and related signage could be added to each street name and which provide the missing context about the individuals in question. Certainly, by making this a requirement of new and revised names that are formally adopted, meaningful progress toward decoloniality can be made.

Pre-Approve Names

An influential tool that some communities have adopted is a list of pre-approved names. For some municipalities, this takes the shape of a database, whereas for others, it may be a list that is maintained on its website or even kept offline. Consultation with the public is a key step for pre-approved names because citizens are a rich source of local knowledge that can inspire greater diversity in naming.

Pre-approved names, when posted publicly, provide examples of the sorts of names the community wishes to use moving forward.

This strategy builds upon the naming policy's values statement and solves two problems relating to monitoring and planning. First, the municipality can better monitor how new or revised names reflect its demographic composition because the process for approving future names will involve some explanation and justification about the meaning and worthiness of the proposed name, particularly for commemorative ones. This step provides an important opportunity for the municipality to collect the demographic data of proposed names and to only approve ones that help it meet its naming priorities. Second, when communities experience rapid growth, the demand to name subdivisions, not to mention the streets and parks they will contain, can outpace the community's capacity for awarding names. For this reason, many municipalities rely upon the developer to devise names, which compromises the community's ability to shape its identity through names, on the one hand, and to meet its naming priorities as stated in its values statement, on the other. To this end, a pre-approved list provides the developer or municipal staff with ready solutions when neighborhoods expand. By regularly issuing a call to the public for name proposals, moreover, the list can be replenished. And finally, a pre-approved list gives the community the opportunity to prioritize certain themes—for instance, a new sub-division whose names will celebrate women or Indigenous people from the area.

Education and Outreach

An important pathway toward decoloniality is education, and this can be achieved both within educational institutions and through partnerships between schools and naming authorities at any level of government. By introducing place name–related activities within the classroom, the research capacity of students at the intersection of their connection to the community will be harnessed as a powerful tool for change. Doing so builds upon other forms of commemoration, such as statues and architecture, with which students have already grown familiar through class trips and other travels.[21] By looking at street names no differently, students and the broader public will see how quotidian and widespread public commemoration has become and the multifarious ways that it reflects (or not) our identities.[22] This may take many forms and here I explore the following three approaches:

Curriculum Development and Experiential Learning

Schools are strategic sites for addressing historical wrongs. These opportunities rest in the hands of teachers, who design activities and implement curriculum; students, who achieve educational milestones linked to curricular and program requirements; and education policy makers, who steer values and learning outcomes in particular directions.[23] Teachers are well-positioned to implement naming initiatives within their classes. The research project for Sackville, for example, was undertaken by about 50 undergraduate students for a second-year university course about how maps intersect with identity. As a group project, it substituted a substantial final essay and for it they researched and analyzed the town's names, culminating in a report for the town.

In addition to learning about place names, policy, and the community's history, students also developed professional skills and produced a deliverable that extended beyond the classroom, which enhances a sense of contribution, citizenship, and achievement because, unlike a final essay, others beyond the classroom benefited from the project. By engaging students in critical reflection about commemorative naming, they grow aware of the indoctrination many Canadians experience as youths while their identities are being shaped and that often goes unquestioned. This emphasis on education over indoctrination comprises an important means of achieving buy-in for decolonized naming practices. It also chafes against the institutionalized ways that we learn about our identities, for instance through federal and provincial heritage programs, as well as public memory rituals, such as Remembrance Day, and their relation to curriculum development at the elementary and secondary levels in particular.[24]

Involving university students in naming practices can also be the subject matter of a course, which presents an important opportunity to develop their skills in community-focused ways. For instance, after gaining approvals from the university's research ethics board as well as forming a partnership between my university and a nearby First Nation's Chief and Council, in the near future, I will engage Indigenous university students at their community's reservation to analyze the community's street names, all of which are rendered in English and date from the introduction of government postal services in the community. The project translates into a university-level course that prioritizes experiential learning and community engagement as vectors for learning. Experiential learning is a powerful means through which values and context for the past can be explored, particularly in an intergenerational setting such as an Indigenous community, but also in communities more broadly.[25] It also activates students' attraction to activism by increasing their awareness about the politics behind public history and commemoration.[26]

For this course, the students will analyze what the names mean relative to their locations and who and what is represented by them. Working in partnership with the reservation's educational authority and band council office, the university students will consult Elders throughout the community about traditional and contemporary names and related stories that may not be reflected by street and park names. They will launch a social media campaign that consults the broader community about names, as well as request suggestions for new names. The students will also research place name policy and best practices within the community and other Indigenous ones and prepare and present a professional report to Chief and Council as a capstone assignment for which they will earn credit toward their degrees. In parallel, they will develop and launch community-facing outputs to raise awareness about the research and make it accessible to all segments of the community. The course also dovetails in important ways with some of the community's signature priorities, including language revitalization. The format of this course resembles the character of the class project undertaken for Sackville, which demonstrates ways that this type of activity can be scaled up or down according to the resources on hand.

A similar project might be undertaken by primary and secondary schools. As Lindsay Gibson has argued, commemoration controversies and education extending from them should be an essential element of both teaching and learning at these levels.[27] The focus could be narrowed to include the neighborhood in which a school is located or students could be asked to research the names of the streets they walk along as they go to school, or upon which they reside. After students research the meaning of names, broader discussion about commemoration can be led by the teacher, who can draw upon recent news and events that complicate the naming status quo. Students can also contribute to an extracurricular deliverable managed by either a university or the municipality: they can complete a simple database entry that prompts them to indicate any identity-related elements of the name so to produce a powerful dataset that both researchers and the municipality can use for place name and community research. By giving students the opportunity to research and perform data entry, researchers and government staff responsible for place naming will build an important resource for naming without having to find the resources to pay for it, all while enhancing civic engagement and developing students' applied skillsets.

Research Capacity and Training

As this approach to curriculum development demonstrates, building research capacity in place name science will help naming authorities meet their decolonial objectives, whose offices are typically understaffed and not designed for research. Other benefits for building research capacity include enhancing the ability and confidence of citizens to engage with complex questions relating to identity in the past along with a willingness to dialogue today's values with historical commemorations through the knowledge that new tools make accessible to everybody.[28]

For example, the synthesis of municipal policy at the national level and a related database explored earlier in this chapter was a project undertaken to answer a gap in research and naming resources when it comes to understanding local naming priorities and how they may be influenced by policy, when it exists. Funded through a research grant, the students who worked on this project were trained and remunerated throughout its duration. First, four undergraduate students searched the websites of all Canadian municipalities for place name policy; they then imported the policy into the project's database. Subsequently, they contacted municipalities to make further inquiries and to see if they would like to receive the results of our study. Four graduate students created a searchable database where the policy was housed and went about analysing it for identity-related questions that yielded the insights referred to earlier in this chapter. We then co-wrote, visualized, and designed the final report, alongside a model policy for inclusive naming, which was published by Canada's national granting council, making them co-authors on a deliverable that has gone on to have influence at the municipal level where our recommendations for policy have caught the attention of place naming authorities.

These research capacity and training activities create opportunities to develop talent while scaffolding undergraduate and graduate students in positions of leadership with defined responsibilities. Building research capacity will help overcome resource scarcity at all levels of government because they produce the tools needed to understand the state of existing naming practices. As scholars have noted, capacity building in these areas generates various forms of professional capital—from publications and cultural acumen to partnerships with non-academic groups and policy writing.[29]

Connecting with the Public

A key component of these projects is the outreach element through which the general public is consulted, potential names solicited from them, and research-related conclusions is broadly disseminated.[30] These projects also build tools that directly benefit the public, such as a name bank populated with names that the community has proposed, which can regularly nourish the community's list of pre-approved names. But also, understanding the demographic composition of existing names will help build buy-in from the community for embracing the municipality's naming priorities for inclusion and decolonization. Therefore, when it comes to education and outreach, these three objectives—knowing who is already celebrated by a name, comparing the status quo against current demographic information for the community, and facilitating community contributions—raise public awareness about the need for change.

At the same time, past experiences of colonial violence can radiate intersectionally across society.[31] By raising awareness of a racialized name that problematically refers to one group, another group who has experienced similar naming practices can feel seen and share in the restorative actions that may follow. The removal of statues has attracted perhaps more attention in recent years; by grafting the rationale for those removals upon names, greater awareness and support for de-commemoration may be achieved.[32] This approach, importantly, builds buy-in through other means: by finding ways of uniting all of us impacted by past naming decisions through allyship, we are more likely to participate in name restitution and support updating policy and related naming practices.

Place Stories

Storytelling is a powerful means of gathering place name data alongside thinking about place names using decolonial approaches. As Bryan Smith points out, place stories narrativize our connection and interaction with a place.[33] Processes of relationality play an essential role in sharing Indigenous histories and cultural knowledge; they also are a fundamental means for intergenerational knowledge transfer in many families.[34] Many colonial names become so familiar that we fail to see them as visible markers of colonial violence. Once colonial names are storified, they become more visible, tangible, and thus sites of critical engagement with the past.

Sharing knowledge in this way cultivates community interest in place names through place stories and intergenerational storytelling that elucidate local names absent from government databases and common sources of cartographic information, such as Google Maps.[35] Place stories are a powerful technology for understanding a community's attachment to or embrace of a name and can compel naming decision-makers to think more deeply about the importance of naming.[36] For some people, Elsipogtog in Mi'kmaw means either River of Fire or Place Where the Sun Rises and Reflects on the Water, which describes how the eastern sun glows red as it rises over the river.[37] Both rich and interconnected meanings inform us of the place where we paddle while making our way down this river and example what I have called elsewhere "sitting in place," which builds upon Keith Basso's 1996 book about Apache names, *Wisdom Sits in Places*.[38] By being in a place and interacting with all that it contains, we learn more about the roots of a place's name as well as its profound attachment to the place. In contrast, Ryerson University in Toronto (Ontario) examined the story behind its namesake, associated with racialized violence, and decided to unmoor itself from that legacy by changing its name.

Storytelling of this nature disrupts how we typically think of place identity. The static nature of place names and emblems in the twenty-first century is largely a settler-colonial and European construct, on the one hand, but also text has been weaponized as a technology of colonization, on the other.[39] Decolonized and precolonial names in the Americas tend to be experiential, descriptive, and scientific.[40] They can also be expressed in other modalities (i.e., through images and textures).[41] These qualities allow the person to know where they are and what they can do or find in a place, underlying which are multivalent layers of narrative, as the story of Elsipogtog demonstrates. For many Indigenous and settler people, these qualities allow the person to know where they are and what they can do or find in a place, which makes many decolonized names story-based.

Multivocality is a powerful philosophy that undergirds democratic processes throughout Indigenous and settler Canada. On the one hand, concepts such as the *vox populi* suggest that place stories might be shaped by the most recurring narrative threads, perhaps risking the discarding of outlier narratives, in ways we might compare to the first-past-the-post electoral system where the person with the most votes wins. On the other, multivocality can cherish all voices, and especially the outliers, to include everyone's stories, which then link to their identities and experiences of a place, in ways that we might compare to the single-transferable-vote electoral system.

As scholars have established, multivocality empowers marginalized and underrepresented groups to share experiences that may not be at first mainstream, thus supporting the inclusion of alternative understandings of the past.[42] Doing so unsettles the colonial, nationalistic, and imperial narrative threads that often rear their heads in mainstream naming practices. Rooting ourselves to a place and then hearing others' stories about it, moreover, deepens our understanding of its significance and how its name informs our identities.[43] By linking place naming to a value that most people cherish—democratic inclusion and consultative decision-making processes—we transform policy, procedure, and naming priorities.

Place stories draw upon both the name and the name's meaning; they are a powerful technology for intergenerational knowledge transmission and for teaching values that can cut across language and cultural divides while educating others about linguistic and cultural facets of the place with which they may be unaware. By linking place stories to the map, we can foster rich, intergenerational exchanges while involving other data points collected through this form of research.[44] Place stories are therefore a critical educational strategy that will generate broader interest and participation in place naming as well as buy-in for sharing place knowledge, and this form of storytelling appeals to a wide segment of society, including children.

Bringing place stories into dialogue with policy can influence decision-makers to consider alternative pathways to naming while disrupting how we think about place name authority. By empowering broad audiences to contribute their rich knowledge about a place and through a multivocal approach to understanding a place's history and name, a new form of place name authority emerges at the local level that can help guide new policy and decision-making processes at higher levels. Place stories can be pursued in two key ways:

Community-Engaged Learning

Intergenerational place stories and related illustrated books created by students offer an influential way of expanding toponymic knowledge at the local level.[45] A curricular module that regional cultural associations, school boards, religious groups, among others, can adopt or support will stimulate the intergenerational exchanges that undergird language revitalization, the transfer of values from one generation to the next, knowledge of the past to inform the present, and so on.

For example, as part of their studies, youth can ask an Auntie or Elder in their life to tell them a story about a place that reveals the meaning of its name that they believe is broadly used or was known to exist. Students will then prepare a storybook that illustrates what they learned using a combination of text and image, or even a digital story that can be circulated to broader audiences, should their Elders consent to sharing their knowledge.[46] These stories can then be shared with classmates, family, and neighbors; they can be kept by the knowledge keeper, who may view the knowledge as not suitable for broader audiences; they can be performed within the school, archived in the school's library, or preserved as a dataset that links into the data tools created through education and outreach explored earlier in this chapter and which go on to inform community place knowledge activities.

Evidently, community-engaged learning offers an important opportunity for students to give back to their communities in ways that increase students' awareness of the positive impacts their contributions can have. At the same time, this approach draws upon sources of knowledge often excluded from formal educational settings. Community-engaged research should aim for co-creation and co-development between researchers, students, and the community, which requires consultation when communities such as Elsipogtog become involved. Municipal staff can use the principles of community-engaged learning to inform their own planning practices as well. One Australian study concluded that urban planning, when it solicits and responds to place stories beyond settler-colonial ones, will "activate histories of place that the future can continue to encounter, rather than repressing, prescribing or fixing them as past."[47] Another Latin American study observed that by rethinking how urban planners interpret local stories, and when Indigenous epistemology is embraced through the vehicles of the story and storyteller, both storytelling and planning can be decolonized, which has the knock-on effect in this case of ensuring that names and naming practices, too, are decolonized.[48]

Outreach

Outreach to non-academic audiences can take various forms. Community-engaged learning necessarily involves a form of community outreach based, ideally, on reciprocity. Mobilizing research results beyond the academy examples a second means of outreach community included in our research dissemination plans; this might occur through editorials or media interviews that are intended to stimulate community interest in naming. Creating an outreach campaign on social media networks and through other fora to draw broad audiences to examine their places on the map is another way to encourage people to provide alternative spellings, names, name meanings and histories, and parallel names in other languages.[49]

In this way, outreach can offer a steady conduit of new place name data for the researcher while the research team and its community partners pursue meaningful change with naming authorities, binding them in a reciprocal relationship.

Failure to consult communities about the meanings and stories behind their names highlights the fact that the authorities of place knowledge do not necessarily work for the government or the academy, the former of which will be responsible for implementing new or changing names. As we have seen, many outdated policies recommend naming places using Indigenous terms or names, yet they lack any process or established relationship through which consulting these groups should be undertaken and the supports to make the process inclusive and reciprocal. This situation has led to renaming mishaps, exampled by the recent attempt to change Red Indian Lake (Newfoundland) to the Mi'kmaw name, Wantoqo'ti Qospem (Peaceful Lake), without broadly consulting Indigenous and settler residents in the area. Many Mi'kmaw community members objected because the lake is in the traditional territory of the Beothuk, whose experience there after colonization was anything but peaceful. This multivocal response to an erroneous name change that was implemented after only consulting a band council chief revealed to provincial authorities the story behind the lake and thus resulted in influencing, in a multivocal fashion, which name was ultimately chosen (Lake Beothuk).

Conclusion

By creating tools, including datasets, along with policy and best practices for naming that democratically involve everybody, inclusive and decolonial approaches to place naming will reach broad target audiences of varying ages, ethnicities, language backgrounds, organizations, and geographies at local, regional, national, and international levels.[50] At the heart of these tools is the recognition of place stories as a technology for including marginalized place knowledge, education and outreach to involve broader audiences in naming practices, and policy designed to guide decolonial and inclusive place naming practices. The infrastructure exists in Canada to support these tools, whether in the form of municipal, provincial, and territorial or national place name data resources. The resources—when more people are involved in naming—may well be limitless.

Notes

1 R. Rose-Redwood, I. Baird, E. Palonen, and C. Rose-Redwood, "Monumentality, Memoryscapes, and the Politics of Place," *ACME: An International Journal for Critical Geographies* 21.5 (2022), pp. 448–467; Jennifer Wigglesworth, "The Cultural Politics of Naming Outdoor Rock Climbing Routes," *Annals of Leisure Research* 25.5 (2022), pp. 597–620; R. Rose-Redwood, R., D. H. Alderman, and M. Azaryahu, eds., *The Political Life of Urban Streetscapes: Naming, Politics, and Place* (New York: Routledge, 2018); Jodi A. Byrd, *The Transit of Empire: Indigenous Critiques of*

Colonialism (Minneapolis: University of Minnesota Press, 2011); D. L. Berg and J. Vuolteenaho, eds., *Critical Toponymies: The Contested Politics of Place Naming* (Burlington, VT: Ashgate, 2009).

2 Lauren Beck, *Canada's Place Names and How to Change Them* (Montreal: Concordia University Press, 2022); J. R. Akerman, ed., *The Imperial Map: Cartography and the Mastery of Empire* (Chicago: University of Chicago Press, 2009); R. Rose-Redwood, N. B. Barnd, A. H. Lucchesi, S. Dias, and W. Patrick, "Decolonizing the Map: Recentering Indigenous Mappings," *Cartographica* 55.3 (2020), pp. 151–162; J. R. Akerman, ed., *Decolonizing the Map: Cartography from Colony to Nation* (Chicago: University of Chicago Press, 2017).

3 See Geographical Names Board of Canada, *Guiding Principles* (Ottawa: Natural Resources Canada, 2011).

4 See E. R. Jenkins, P. E. Raper, and L. A. Möller, *Changing Place Names* (Durban: Indicator Press, 2016), along with S. Cashin, *Place, Not Race: A New Vision of Opportunity in America* (Boston, MA: Beacon Press, 2014); Dorcas Zuvalinyenga and A. Libert, "An Analysis of the Relationship between Toponyms and a Variety of Historical and Cultural Specificities in the Discursive Construction of Identity in a Regional Town in Zimbabwe," *Onoma: Journal of the International Council of Onomastic Sciences* 54 (2019), pp. 77–98.

5 S. Castro-Gómez, *La poscolonialidad explicada a los niños* (Popayán: Editorial Universidad del Cauca, 2005); Enrique Dussel, *The Invention of the Americas: Eclipse of 'The Other' and the Myth of Modernity* (New York: Continuum, 1995); C. Dwyer and C. Bressey, eds., *New Geographies of Race and Racism* (Burlington, VT: Ashgate, 2008); G. L. Eades, ed., *Maps and Memes: Redrawing Culture, Place, and Identity in Indigenous Communities* (Montreal: McGill-Queens University Press, 2015).

6 A. Lyall, H. Nelson, D. Rosenblum, and M. Turin, "K̲a̲k̲otłatła̲no'x̲w̓ x̲a k̲wak̲wax̲'mas: Documenting and Reclaiming Plant Names and Words in Kwakwala on Canada's West Coast," *Language Documentation and Conservation* 13 (2019), pp. 401–425; G. L. Eades, *The Geography of Names: Indigenous to Post-Foundational* (New York: Routledge, 2017); J. Episkenew, *Taking Back Our Spirits: Indigenous Literature, Public Policy, and Healing* (Winnipeg: University of Manitoba Press, 2009); L. Irwin, *Coming Down from above: Prophecy, Resistance, and Renewal in Native American Religions* (Norman: University of Oklahoma Press, 2008).

7 Jennifer Wigglesworth, "Renaming Routes: A Tale of Oppression and Resistance in Outdoor Sport Climbing," in *Gender, Politics, and Mountaineering: A Mountain Still to Climb*, eds. J. Hall, E. Babcock, and Z. Arner (New York: Palgrave Macmillan, 2023), pp. 129–150; L. Bigon and D. Zuvalinyenga, "Urban Pulse-Gendered Urban Toponyms in the Global South: A Time for De-Colonization?" *Urban Geography* 42.2 (2020), pp. 226–239; D. G. Cole and I. Sutton, "A Cartographic History of Indian-White Government Relations over the Past 400 Years," *American Indian Culture and Research Journal* 37.1 (2013), pp. 5–77.

8 In a New Zealand context, see N. Lindsay and R. Kearns, "The New Zealand Geographic Board and the Contested Nature of Place Names in New Zealand," in *Place Naming, Identities and Geography: Critical Perspectives in a Globalizing and Standardizing World*, ed. G. O'Reilly (Berlin: Springer, 2023), pp. 241–257; R. Kearns, "Why We Should Embrace Māori Renaming," *ideasroom*, 2023.

9 Frédéric Giraut and M. Houssay-Holzschuch, eds., *The Politics of Place Naming: Naming the World* (London: Wiley, 2022); Z. Grossman, *Unlikely Alliances: Native Nations and White Communities Join to Defend Rural Lands* (Seattle: University of Washington, 2017); José Rabasa, *Franciscans and Dominicans under the Gaze of a Tlacuilo: Plural-World Dwelling in an Indian Pictorial Codex* (Berkeley: University of California, 1998).

10 See M. M. Kretzer and R. H. Kaschula, "Language Policy and Linguistic Landscapes at Schools in South Africa," *International Journal of Multilingualism* 18.1 (2021), pp. 105–127; C. W. Post and D. H. Alderman, "'Wiping New Berlin off the Map': Political Economy and the De-Germanisation of the Toponymic Landscape in First World War USA," *AreaREA* 46.1 (2014), pp. 83–91; D. H. Alderman and J. Inwood, "Street Naming and the Politics of Belonging: Spatial Injustices in the Toponymic Commemoration of Martin Luther King Jr.," *Social & Cultural Geography* 14.2 (2013), pp. 211–233; T. Banivanua Mar and P. Edmonds, eds., *Making Settler Colonial Space: Perspectives on Race, Place and Identity* (New York: Palgrave Macmillan, 2010); H. L. Gates Jr., *Tradition and the Black Atlantic: Critical Theory in the African Diaspora* (New York: BasicCivitas, 2010); Paul Gilroy, *The Black Atlantic: Modernity and Double Consciousness* (Cambridge, MA: Harvard University Press, 1993).

11 W. Kirkconnell, *Canadian Toponymy and the Cultural Stratification of Canada* (Winnipeg: Ukrainian Free Academy of Sciences, 1954). Also see L. Kostanski and G. Puzey, eds., *Names and Naming: People, Places, Perceptions and Power* (Blue Ridge Summit, PA: Multilingual Matters, 2016); J. M. O'Brien, *Firsting and Lasting: Writing Indians Out of Existence in New England* (Minneapolis: University of Minnesota Press, 2010); Stephanie Lewthwaite, *Race, Place, and Reform in Mexican Los Angeles: A Transnational Perspective, 1890–1940* (Tucson: University of Arizona Press, 2009); Mark Monmonier, *From Squaw Tit to Whorehouse Meadow: How Maps Name, Claim, and Inflame* (Chicago: University of Chicago Press, 2006).

12 See L. D. Berg and R. A. Kearns, "Naming as Norming: 'Race,' Gender and the Identity Politics of Naming Places in Aotearoa/New Zealand," in *Critical Toponymies: The Contested Politics of Place Naming*, eds. L. D. Berg and J. Vuolteenaho (Burlington, VT: Ashgate, 2009), pp. 19–51.

13 M. Doucette, "Listening for Awoodet: Searching for Demasduit in a Lost Beothuk Vocabulary," *The Independent*, March 8, 2023; D. Igdoura, "An Examination of Settler Colonialism in Canada's Legal Institutions: 1492 Land Back Lane," *Aletheia* 1.2 (2021), pp. 39–46; Rose-Redwood et al., "Decolonizing the Map," pp. 151–162; D. Rück and C. Gray, "Reclaiming Indigenous Place Names," *Yellowhead Institute Policy Brief*, 2019; R. Sockbeson, "Honored and Thriving: The Squaw Law and Eradication of Offensive State Place-Names," *American Indian Culture and Research Journal* 40.2 (2016), pp. 123–138; Marc Beaulieu, "An Analysis of Cobo's Hagiotoponyms of South American and Related Indigenous Toponyms," *Onomastica Canadiana* 93 (2015), pp. 1–12; Glen Coulthard, *Red Skin, White Masks: Rejecting the Colonial Politics of Recognition* (Vancouver: UBC Press, 2014); Barbara E. Mundy, "Place-Names in Mexico-Tenochtitlan," *Ethnohistory* 61.2 (2014), pp. 329–355; C. Bartlett, M. Marshall, and A. Marshall, "Two-Eyed Seeing and Other Lessons Learned within a Co-Learning Journey of Bringing Together Indigenous and Mainstream Knowledges and Ways of Knowing," *Journal of Environmental Studies and Sciences* 2.4 (2012), pp. 331–340; Z. Grossman and A. Parker, *Asserting Native Resilience: Pacific Rim Indigenous Nations Face the Climate Crisis* (Corvalis: Oregon State University Press, 2012); G. Tibón, "Mexico: The Rabbit's Navel," *Names: A Journal of Onomastics* 39.3 (1991), pp. 257–260.

14 P. Jordan and P. Woodman, eds., *Place-Name Changes: Proceedings of the Symposium in Rome, 17–18 November 2014* (Hamburg: Verlag Dr. Kovac, 2016).

15 C. Desbiens, I. Hirt, and B. Collignon, "Introduction to the Special Section on Indigenous Spatial Capital: Incorporating First Peoples' Knowledges, Places, and Relations into Mapping Processes," *The Canadian Geographer/Le Géographe canadien* 64 (2020), pp. 4–9.

16 See J. Schillo and M. Turin, "Type Right: Examining the Underlying Causes of Common Typeface and Font Errors for Indigenous Orthographies, and a Possible Path Forward," *Language Documentation and Conservation* 16 (2022), pp. 364–398; M. L. Berman, R. Mostern, and H. Southall, eds., *Placing Names: Enriching and Integrating Gazetteers* (Bloomington: Indiana University Press, 2016).

17 CBC (Canada Broadcasting Corporation), "Vancouver City Councillor Wants City Signage to Reflect Region's Diversity," *CBCNews*, March 28, 2017, p. 4; Calgary Aboriginal Urban Affairs Committee, *White Goose Flying: A Report to Calgary City Council on the Indian Residential School Truth and Reconciliation Calls to Action 2016* (Calgary: The City of Calgary, 2016).

18 https://placenames.cemvc.ca/.

19 L. Beck, V. MacBeath, A. Walker, and J. Yoston, *Identity in Canada's Place Name Policy: A Knowledge Synthesis* (Ottawa: Social Sciences and Humanities Research Council, 2023).

20 L. Beck et al., *Our Names, Our Home: Report on the Town of Sackville's Street Names and Monuments* (Sackville, NB: Mount Allison University, 2022).

21 Penney Clark and Alan Sears, "Points of Convergence: Public Commemorative Art and the Teaching of History" in *The Arts and the Teaching of History: Historical F(r)ictions*, eds. Penney Clark and Alan Sears (Cham: Palgrave Macmillan, 2020), pp. 165–197.

22 Penney Clark and Alan Sears, "Engaging the Fray: Preparing Teachers and Students for Critical Encounters with the Past," in *The Arts and the Teaching of History: Historical F(r)ictions*, eds. Penney Clark and Alan Sears (Cham: Palgrave Macmillan, 2020), pp. 199–242. Also see David Neufeld, "Parks Canada and the Commemoration of the North: History and Heritage," in *Northern Visions: New Perspectives on the North in Canadian History*, eds. Kerry Abel and Ken Coates (Toronto: University of Toronto Press, 2001), pp. 45–75 and Carol Mutch and Vanessa Cameron-Lewis, "Teaching about War Today: An Analysis of Curriculum Resources Produced for New Zealand's Commemoration of the First World War," *Citizenship, Social and Economics Education* 16.3 (2017), pp. 184–198.

23 James Miles, "Redressing Historical Wrongs or Replicating Settler Colonialism? Social Studies Curriculum Reform in Canada," in *Historical Justice and History Education*, eds. Matilda Keynes, Henrik Åström Elmersjö, Daniel Lindmark, and Björn Norlin (Cham: Palgrave Macmillan, 2021), pp. 249–267.

24 On this shift toward education about commemorative practices, see Alan Sears, "Respectful Commemoration," in *Learning to Inquire in Social Studies: An Anthology for Elementary Teachers*, eds. Roland Case and Penney Clark (Vancouver: The Critical Thinking Consortium, 2021), pp. 97–102.

25 This is explored in the context of the Holocaust by Ruth Dorot, Shlomit Ben-Ishay, and Nitza Davidovitch, "The Contribution of Monuments to Educating about Holocaust Commemoration in Israel," *International Education Studies* 14.7 (2021), pp. 80–90.

26 Erika Dyck, "Doing History that Matters: Going Public and Activating Voices as a Form of Historical Activism," *Journal of the History of the Behavioural Sciences* 57.1 (2021), pp. 75–86.

27 Lindsay Gibson, "The Case for Commemoration Controversies in Canadian History Education," *Canadian Journal of Education* 44.2 (2021), pp. 434–465.

28 Nicola Gauld and Ian Grosvenor, "The Role of Commemoration in History and Heritage: The Legacy of the World War One Engagement Centres," in *Historical Justice and History Education*, eds. Matilda Keynes, Henrik Åström Elmersjö, Daniel Lindmark, and Björn Norlin (Cham: Palgrave Macmillan, 2021), pp. 153–175.

29 Chris Holligan, "Becoming a Researcher: Forms of Capital Associated with 'Research Capacity' Trajectories of Young British Social Anthropologists," *International Journal of Qualitative Studies in Education* 28.10 (2015), pp. 1249–1270.
30 Gauld and Grosvenor, "The Role of Commemoration in History and Heritage," pp. 153–175.
31 This is explored in terms of shared experiences of genocide by Jordan Stranger-Ross and Lynne Marks, "Remembering the Holocaust in a Settler Colonial City: The Case of Victoria, Canada," *History & Memory* 34.1 (2022), pp. 100–132.
32 See the chapters in Sarah Gensburger and Jenny Wüstenberg, eds., *De-Commemoration: Removing Statues and Renaming Places* (New York: Berghahn, 2024).
33 Bryan Smith, "Cartographies of Colonial Commemoration: Critical Toponymy and Historical Geographies in Toronto," *Journal of the Canadian Association for Curriculum Studies* 15.2 (2017), pp. 34–47.
34 Cody Groat and Kim Anderson, "Holding Place: Resistance, Reframing, and Relationality in the Representation of Indigenous History," *Canadian Historical Review* 102.3 (2021), pp. 465–484.
35 D. H. Alderman and R. Rose-Redwood, "The Classroom as 'Toponymic Workspace': Towards a Critical Pedagogy of Campus Place Renaming," *Journal of Geography in Higher Education* 44.1 (2020), pp. 124–141; G. Towers et al., "Diversifying for Sustainability: Repurposing a Targeted Pilot Faculty Mentoring Program," in *10th Annual Mentoring Conference Proceedings: A Decade of Cultivating an Inclusive Mentoring Community*, eds. N. Dominguez, B. Berkeley, N. Barka, T. Chrisman, B. Kelley, and E. Westfall (Albuquerque: University of New Mexico Press, 2017), pp. 818–823; D. Hunt, "Nikîkîwân 1: Contesting Settler Colonial Archives through Indigenous Oral History," *Canadian Literature* 230/231 (2016), pp. 25–42; D. L. Moore, *That Dream Shall Have a Name: Native Americans Rewriting America* (Lincoln: University of Nebraska Press, 2013); N. Kaufman, *Place, Race, and Story: Essays on the Past and Future of Historic Preservation* (New York: Routledge, 2009).
36 C. Aporta, I. Kritsch, A. Andre, K. Benson, K., S. Snowshoe, W. Firth, and D. Carry, "The Gwich'in Atlas: Place Names, Maps, and Narratives," in *Developments in the Theory and Practice of Cybercartography*, ed. D. R. F. Taylor (Amsterdam: Elsevier, 2001), pp. 229–244.
37 M. W. McDonald and A. Julian, "Elsipogtog First Nation," 2022, https://www.apcfnc.ca/wp-content/uploads/2020/06/Elsipogtog_First_Nation_article.pdf.
38 Keith H. Basso, *Wisdom Sits in Places: Landscape and Language among the Western Apache* (Albuquerque: University of New Mexico Press, 1996).
39 J. Rappaport and T. Cummins, *Beyond the Lettered City: Indigenous Literacies in the Andes* (Durham, NC: Duke University Press, 2012); G. Vizenor, *Manifest Manners: Narratives on Postindian Survivance* (Lincoln: University of Nebraska Press, 1999); Linda Tuhiwai Smith, *Decolonizing Methodologies: Research and Indigenous Peoples* (New York: Zed Books, 1999); Walter J. Ong, *Orality and Literacy: The Technologizing of the Word* (New York: Routledge, 1982).
40 M. R. Goeman, *Mark My Words: Native Women Mapping Our Nations* (Minneapolis: University of Minnesota Press, 2013); R. Wall Kimmerer, *Braiding Sweetgrass: Indigenous Wisdom, Scientific Knowledge, and the Teachings of Plants* (Minneapolis, MN: Milkweed, 2013); M. R. Goeman, "Notes toward a Native Feminism's Spatial Practice," *Wicazo Sa Review* 24.2 (2009), pp. 169–187; C. Evans, J. Pettigrew, Y. Kromchain Tamu, and M. Turin, *Grounding Knowledge/Walking Land: Archaeological Research and Ethno-Historical Identity in Central Nepal* (Cambridge: McDonald Institute for Archaeological Research, 2009).

41 D. Y. Arnold and J. de Dios Yapita, *The Metamorphosis of Heads: Textual Struggles, Education, and Land in the Andes* (Pittsburgh: University of Pittsburgh Press, 2006); Barbara E. Mundy, *The Mapping of New Spain: Indigenous Cartography and the Maps of the* relaciones geográficas (Chicago: University of Chicago Press, 2000); Elizabeth Boone and Walter D. Mignolo, eds., *Writing without Words: Alternative Literacies in Mesoamerica & the Andes* (Durham, NC: Duke University Press, 1994).
42 See the chapters in Junko Habu, Clare Fawcett, and John M. Matsunaga, eds., *Evaluating Multiple Narratives: Beyond Nationalist, Colonialist, Imperialist Archaeologies* (New York: Springer, 2008).
43 This is explored throughout Katherine Meizel, *Multivocality: Singing on the Borders of Identity* (New York: Oxford University Press, 2020).
44 Juan Sánchez, *Uranium* (Mexico City: Cactus del Viento, 2003); G. Brotherston, *Book of the Fourth World: Reading the Native Americas through Their Literature* (New York: Cambridge University Press, 1992); K. Smith, "Like the Chameleon Who Takes on the Colors of the Hills: Indigeneity as Patrimony and Performance in Coastal Ecuador," *Journal of Latin American and Caribbean Anthropology* 21.1 (2016), pp. 19–38; F. C. Shipek and R. Pinto Robertson, eds., *The Autobiography of Deflina Guero, A Diegueño Indian, as Told to Florence C. Shipek, Interpreter Rosalie Pinto Robertson* (Los Angeles, CA: Dawson's Book Shop, 1968).
45 Also see S. Dunn and V. Vitale, "Linking Text and Maps: Annotation as a Critical Tool for Teaching in the Spatial Humanities," *Literary Geographies* 7.2 (2021), pp. 292–310; T. Adcock, G. Jackson, and J. Sánchez, "Water Protectors: Stories on Inhabited Territories," in *Reading in Race, Ethnicity, Immigration and Minority Relations*, ed. H. Jimenez (San Diego, CA: Cognella Academic Publishing, 2017), pp. 254–272.
46 See, for instance, Christina Fisanick and Robert O Stakeley, *Digital Storytelling as Public History: A Guidebook for Educators* (New York: Routledge, 2020).
47 Emily Potter, "Contesting Imaginaries in the Australian City: Urban Planning, Public Storytelling and the Implications for Climate Change," *Urban Studies* 57.7 (2020), pp. 1536–1552: 1548.
48 Catalina Ortiz, "Storytelling Otherwise: Decolonising Storytelling in Planning," *Planning Theory* 22.2 (2023), pp. 177–200.
49 Along the lines of R. Kearns and L. D. Berg, "Proclaiming Place: Towards a Geography of Place Name Pronunciation," *Social & Cultural Geography* 3.3 (2002), pp. 283–302 and not Felipe Fernández-Armesto, *Amerigo: The Man Who Gave His Name to America* (New York: Random House, 2007).
50 R. Rose-Redwood, C. Rose-Redwood, D. H. Alderman, and K. Hackett, "The Making of the Campus Namescape: A Comparison of University Naming Policies in Canada and the United States," *The Professional Geographer* 76.3 (2024), pp. 277–288; P. J. Deloria, *Indians in Unexpected Places* (Lawrence: University Press of Kansas, 2004); Walter D. Mignolo, *Local Histories/Global Designs: Coloniality, Subaltern Knowledge, and Border Thinking* (Princeton, NJ: Princeton University Press, 2000).

5 Decolonizing Names and Dynamics of Cross-Cultural Naming Practices Among the Urhobo and Yoruba Ethnic Groups in Nigeria

Joy Aruoture Omoru and Israel Abayomi Saibu

The Urhobo and Yoruba are two ethnic groups in Nigeria with rich cultural heritages and traditions, and the act of naming within these cultures is far more profound than just picking a label. For the Urhobo and Yoruba people, naming is a unique tradition that is much more than simply labeling a child, place, or thing. It is a tradition that reflects Nigerian values and histories, one that expresses their aspirations, makes pronouncements on destiny, and establishes connections to lineage and ancestry. However, the colonial experience left an indelible mark on this practice among the Urhobo and Yoruba people in Nigeria, hence, the discourse on the importance of names and naming practice, such as those of Benson Ohihon (2014), Dikko Muhammad (2023), Akpomuvie and Esiri (2021), Faloju and Fadairo (2020) and others who have attempted to engage with the issue of naming from various perspectives. These scholars contend that names are cultural configurations that hold philosophical, religious, anthropological, artistic, historical, and other significance that transcends linguistic inquiries on the subject. Such linguistic inquiries focus their attention on the lexical-semantic character of names while looking away from decolonization of names among colonized or marginalized communities and do not give adequate attention to the dynamics of the lexical syntactic characteristics of names.

This exploration delves into the world of Urhobo and Yoruba naming practices. It examines how these traditions function within both cultures, considering the meanings imbued in names and the rituals surrounding their selection. It then addresses the concept of decolonizing names. This outcome involves interrogating the influence of colonialism on naming practices in the Urhobo and Yoruba cultures and exploring how these ethnic groups have attempted to reclaim their own agency in this vital cultural aspect. By examining cross-cultural dynamics, the study promotes a deeper understanding of how these communities navigate the tension between preserving tradition and confronting the legacies of colonialism. For example, across the African continent, and in Nigeria, many individuals, such as Chinua Achebe, Obafemi Awolowo, Nnamdi Azikiwe, Ngugi wa Thiong'o, to mention a few, have rejected their English given names to embrace

DOI: 10.4324/9781003414773-6

their Indigenous names. Among the Urhobo, rather than change names, it is common practice to "urhobonise" English names. For instance, the name Godwin is urhobonised as Godini, Beatrice as Ibiatrisi, Hope as Ihopi, Micheal as Imaikẹli, and more. Beyond individual and community reactions to colonial impact, some countries in Africa changed their names after attainment of independence. For example, Ghana during the colonial era was known as the Gold Coast, but after independence changed to Ghana, Zambia and Mali are other examples. These changes are marks of resistance to colonialism, neocolonialism, or a renaissance of cultural identity.

In contemporary postcolonial Nigeria, a survey of the daily newspapers, for instance, will divulge the magnitude at which people change their individual names or their surnames. Out of ten issues of the Nigeria's Daily Newspaper, one will find such announcement as follows:

> Abayomi- I formerly known and addressed as Mr. Edward Abayomi Adekunle of No. 12 Oluodo Street, Ikorodu Lagos, now wish to be known and addressed as Mr. Abayomi Adebambo Adekunle. All former documents remain valid. The general public please take note.[1]

All these are glaring examples of how Africans attach importance to Indigenous African names as a tool of decolonization process or a change of statues among themselves.

Precolonial Nigeria had rich cultural naming practices. During this period, the Indigenous people displayed their rich cultural naming practices dexterity mostly through narrative of significant occurrence and religious inclination. One powerful method by which they effectively encoded and communicated names and naming practices was by the use of oral tradition. Among the Urhobo and the Yoruba, names given to individuals are carefully selected with significant connotation. For example, the Urhobo names Ejuvweghe: I won't be mocked, and Emusherearhe: Let manner fall from heaven, respectively teaches the child that hard work underscores respect and securing a living (food and more). The Yoruba capture the essence of reincarnation and continuity of a family's greatness in the names: Babatunde/Iyatunde: Father is back/Mother is back. It is therefore considered that names reinforce values that influence the behavior, character, and future well-being of a child. As rightly observed by Akpomuvie and Esiri, if a fool is a name given to a child at birth, that child might grow up to be a fool. Correspondingly, a child whose name connotes success or failure might translate for the child in future ending up either being successful or a failure in life.[2] Hence, parents from these study areas do not just give names to their children without taking the child's future into consideration. Furthermore, the recent development of registering the names and date of birth of a child at the birth registry was not in practices in the study areas during the precolonial period. This practice came with the advent of colonialism. Traditionally, among the Urhobo and the Yoruba, it is forbidden to count the number of children that an individual has and naming of children among these ethnic groups are carefully selected.

Be that as it may, naming practices among decolonized people in Africa and, in particular, among the Urhobo and the Yoruba ethnic groups in Nigeria, have greatly been influenced by the following factors: cultural identity, religion, decolonization, immigration and socio-political factor.[3] Literature on slavery and its impacts have critically examined the names African bears before enslavement and how those African names were changed by their colonial masters, as they were transported and made to sojourn in foreign land. In retrospect, some Nigerian's names were deformed. Some of those names were replaced with foreign names, while many others became victims of linguistic mutilation committed by some extraneous forces.[4] The colonization of Nigeria significantly affected Indigenous names.[5] With colonization, Nigeria's Indigenous names lost its original linguistic connotation, historical, value and scientific significance and functions. Some individuals were completely stripped off their Indigenous names and given foreign names by the colonialist.

Significantly, studies on decolonization of names and the dynamics of cross-cultural naming practices in Nigeria, especially among the Urohobo and the Yoruba ethnic groups are not copious, notwithstanding the all-embracing research on naming practice. Studies on the history and origin of names are abundant.[6] Thus, little effort has been spared to address naming practices and decolonization in Nigeria. Hence, the imperative to examine the dynamics of cross-cultural names, naming practices and decolonization of names in postcolonial Nigeria among the two ethnic groups in the study areas. This research seeks to address these intellectual gaps.

Specifically, this research is a humble effort to lend a scholarly voice to the increasing global call for African cultural renaissance, Indigenous naming and practices, preservation of Indigenous African names and decolonization of names among decolonized people in Africa and beyond. Essentially, this research argues that decolonizing names in postcolonial Nigeria among the Urhobo and the Yoruba is a strong strategy or tool for restoring and preserving Indigenous names and the rich cultural heritage and history of the people. The argument on decolonization of names and naming practices as discussed in this research is an unbiased effort in contributing to the wider contemporary decolonization discourse on the African continent. The arguments presented in this research are built on three premises: first, African Indigenous names were distorted and sometimes replaced during the colonial period and in particular during the ignoble slave trade era; second, Nigerian Indigenous names and naming practices, especially among the Urhobo and the Yoruba ethic groups, are ways of encoding and expressing their dynamic cultural heritage, history and world view; and third, decolonization and restoration of Indigenous names among the Urhobo and the Yoruba and invariably in Africa will resort to renaissance and preservation of the rich Indigenous cultural heritage of the study areas and by extension in Africa. Importantly the paper highlights the need to preserve the rich cultural heritage of the study areas through decolonization of name practice among the people.

In lieu of the need and purpose of the study, an understanding of the concepts of culture, Nigerian cultural identity, naming practices, colonialism, and decolonization are fundamental for assessing names among the Urhobo and Yoruba as cultural configurations that serve as tools of decolonization process and identity constructions.

Nigerian Cultural Context

Names are culture specific. In some cultures, though names have meanings and historical records of usage, they do not encode family, communal, or historical narratives. Hence, they come off as mere labels for people, places, and things. Consequently, culture frames the construction, usage, and purposes of names.

Culture can be defined as a group of artifacts, beliefs, languages, symbols, and values and is made manifest through material and nonmaterial culture. Material culture consists of artifacts (material objects), while nonmaterial culture consists of "values, beliefs, symbols, and language that define a society."[7] Language as an element of nonmaterial culture is considered as an aspect of symbol, a nonmaterial tool in culture used to represent (name) things, and which "evoke various reactions and emotions." Naming is done by language and the meaning and interpretation ascribed to names by which things, people, and places are labeled are dependent on cultural consensus. The feature of agreement in culture makes the use of names in most cultural climes to exert considerable influence on the way the people "think and perceive their world."[8] In this case, naming practices in Urhobo and Yoruba languages would impact the people's perception of their world differently from English names which embody English culture. The ability of Indigenous names to encode and express Urhobo and Yoruba cultural phenomena better than other languages is perhaps one of the reasons that foreground the resurgence in the use of Indigenous Urhobo and Yoruba names to label places and people formerly labeled in English language.

Since names play a semantic function in a culture, understanding the naming practice in specific cultures is germane to understanding their world, which comprises their history and experiences. They lack the capacity to convey shared experiences and values over a long period of time. This perspective of culture points out culture's potential to historically archive and transmit human experiences, ideas, and values through its many elements, which include naming practice. This view of culture portrays the Urhobo and Yoruba cultures in which the people store, transmit, and recount their experiences from one generation to another through word of mouth. In these milieus, names are narrative tools, which households and communities employ to symbolically recount events prior to the moment of naming, and to philosophy about the future and life in general.

M. J. Malukeke defines culture as "traditional cultural practices that reflect the values held by members of a community for periods often spanning generations."[9] Such cultural practices and beliefs, according to him, abound in cultural groups and can be harmful or beneficial to members. Harmful cultural practices are beliefs or ideologies and practices that run contrary to both international and national human rights laws. In the same vein, it could be argued that cultural norms and practices of particular cultures that violate the norms and practices of other cultures through modifications of any kind are harmful to their target cultures. The process by which members of the target audience reorganize themselves to handle the new and the former culture prevents them from being rooted in both cultures. This is the case made by Lawino in Okot p'Bitek's "Song of Lawino" (1984), when she cautions that "the pumpkin in the old homestead must not be uprooted!" This proverb evokes imagery of definite death for the pumpkin, since the transplanting process is wrong and harmful. The pumpkin is symbolic of African and Nigerian cultures such as the Urhobo and Yoruba cultures, which suffered enormous harm from the English cultural practices of naming and renaming of people and places using English labels to which the Indigenous people could not relate semantically.

J. A. Banks and C. A. M. Banks agree with Malukeke's definition of culture but remain silent on whether artifacts should be part of the components of culture. Their concept of culture hinges on the notion that the essence of culture is not the artifacts and other tangible elements, but its intangible symbolic, ideational, and other intangible aspects that influence interpretations, meanings and artifacts, and perspectives.[10] In essence, the meaning people attach to the tangible aspects of their society, the ideas they hold, the stories that they weave from it, and how they document it are the essence of their culture. It may therefore be held that tangible culture is expressions of how members of a cultural group express the interpretation and perception of their world. Names play a significant role in this process of expression. Thus, use of names, among the Urhobo and Yoruba, to identify people, places, and things reveals the people's philosophy on life, their approach to meaning, and more. For instance, every name among the people is foregrounded by a profound narrative and philosophy of life formed by events and beliefs surrounding the individual or community at specific times in history. Names such as Akponhwe-Erhivwinejovwo (The world brought death but the spirit world overruled it), among the Urhobo, and Babatunde/Iyatunde (Father/Mother has been reborn) among the Yoruba identify people, reveal aspects of their worldview on life and death and the complexities of the relationship between the physical and spiritual spheres that constitute their worlds. Names therefore often have narrative and historical connotations.

One helpful definition to this study's inquiry into the decolonization of names among the Urhobo and Yoruba is that by Kluckhohn and Kelly which views culture as consisting of "patterns of ways of thinking, feelings, and reactions acquired and transmitted mainly by symbols."[11] The notion of transmission inherent in this definition of culture is in tandem with the oral nature of Urhobo and Yoruba traditions of documenting and transmitting individual and communal achievements and traditional ideas and values through cultural configurations such as names and other cultural symbols. It can therefore be inferred that imposition of one language on another through naming practices, as observed in colonial regimes in Africa, positions the recipient culture in a dilemma of grappling with conflicting ideas and values, and a threat of erosion of erstwhile Indigenous values and ideas. This scenario makes decolonization an imperative, which can be attained only by retransmitting Indigenous ideas and values through a resurgence in the use of Indigenous names and other symbols as is the case of the Urhobo and Yoruba.

Segall, Campbell, and Herskovits opine that an attempt to clarify the concept of culture is not worth the effort. Since most scholars hardly have a consensus on the definition of the concept, he therefore cautions that cultural analysts should focus instead on searching for whatever cultural variable might link with established variations in human behavior. But the question arises on how these variables ought to be determined. In lieu of this, the contention in the concept of culture comes handy, that "culture can be whatever a scholar decides it to be" and researchers only "need to explain exactly how they propose to measure culture in accordance with their conceptualization."[12] In line with these perspectives, the current study's exploration of some conceptualizations of culture has been guided by the task of examining studies on culture that relate to and guide the work to realize its task of researching the discourse on decolonizing names among the Urhobo and Yoruba cultural groups. This approach is similar to that adopted in *The Study of Culture* where the research effort focuses on the study of culture rather than clarifying the concept. The work therefore explores several methods of studying culture from the perspective of several disciplines. By so doing, it reveals that methodology comes off as one of the points of debate on the conceptualization of culture. Again, Segall et al. call for caution in cross-cultural analysis comes handy to help this research bypass the debate and focus on the task at hand by adopting the cultural approach to interpret the names collated for analysis.

The UN System Task Team on Post-2015 UN Development Agenda views culture from a developmental perspective. The team contends that statistics and data in the cultural sector and more underscore that culture is a powerful driver for development because it impacts society socially, economically, and otherwise. Development from culture, they argue, includes "social inclusiveness and rootedness, resilience, innovation, creativity and entrepreneurship for individuals and communities, and the use of local resources, skills, and knowledge."[13] They conclude that naming as a form of cultural expression can be considered a tool in making people rooted in their culture, achieve social inclusiveness, and become more productive to themselves and community.

Nigeria Cultural Identity and Colonialism

The emergence of colonialism brought together over 450 linguistic or ethnic groups with distinct customs, value systems, cultures, religions, politics, and economy under a singular political umbrella named Nigeria.[14] This fusion of Indigenous states ensured a smooth running of the political affairs of Nigeria and exploitation of its natural resources without hitches. This was foregrounded by British colonialist imposition, on the people, of their customs, culture, education, art, language, religion, and value system. Nigerian traditions and cultural identity were portrayed as uncivilized and archaic heritage that must be cast-off and replaced with British culture. The act of replacement had a great mark on naming and renaming of people, places, and things using English names. Where Indigenous names were not changed, they were anglicized. An instance among Urhobo communities is Otujeremi originally named Otughievwe. The latter means the land or home of the Ughievwe people, while the former hold no explicit meaning in local parlance. Globalization further, in no small way, estranged the people from their roots as a result of the impact of information and communication technology. Through the globalized media, people all over the world are being made to look the same, profess the same faith, speak the same language, wear the same type of dresses, enjoy the same type of music, and eat the same type of food.[15] To unravel how the Nigerian people through the two ethnic groups under consideration have reacted to the cultural onslaught in the face of colonialism has necessitated this research.

Decolonization

The term decolonization was first coined by the German economist Moritz Julius Bonn in the 1930s. He used the term to describe former colonies that attained self-governance. Thus, the term evokes several notions such as colonialism, postcolonialism, and cultural/Indigenous identity. Colonialism often conjures the image of the marginal versus superior, while postcolonialism evokes the image of conscientious efforts to reclaim, recreate, and re-portray the agency of the previously marginal group. These concepts are encapsulated in the notion of decolonization when viewed as a moment and process. As a moment, it deals with historical moments when colonialism and its extension of civilization to other cultures foregrounded by the ideological justification of superiority ascribed to the colonial empire's race and culture comes to an end to usher in independent states. The ideology of superiority permeates linguistic, intellectual, social, and cultural practices such as naming of people, places, and things. Naming in most cultures encapsulates the group's expression of their social, intellectual, and linguistic identity. Hence, an extension of colonial civilization to the group in this dimension, as observed among the Yoruba and Urhobo people in Nigeria, is subjugation. However, it is only as a process that decolonization addresses this subjugation.

As a process, decolonization refers to the many-faceted events and procedures employed by independent states to shake off colonial rule and its consequences. This process is of import to the current study. Consequently, while examining the decolonization of names among the Urhobo and Yoruba in Nigeria, the study's use of decolonization does not designate the historic moment of dissolution of empires and the creation of independent states, but it accentuates the process by which colonized people, especially in Nigeria, delegitimize any kind of rule that is experienced as a relationship of subjugation to a power elite. The process of decolonization, in postcolonial discourse, is viewed as a response and resistance to the various cultural knowledge and methods utilized by colonialists to subjugate people from cultures to which they strive to extend their civilization. According to Bert Olivier, this process demands appropriation of any or all knowledge systems that can help autochthonous people to "achieve relative epistemic autonomy . . . for previously unacknowledged and/or unacknowledged . . . traditions."[16] He contends that only this kind of decolonization can resist colonial power because it turns toward the future rather than repeating the past. This form of decolonization is hinged on Frantz Fanon's notion of decolonization. In his *The Wretched of the Earth*, he views decolonization as the creation of a new human being unaffected by the structuring of human beings into racial or cultural compartments. He contends that the creative process of such a people, which will in turn produce a single indivisible humanity, is a "program of complete disorder . . . a historical process . . . it cannot be understood . . . never takes place unnoticed, for it influences individuals and modifies them fundamentally."[17] These perspectives are enshrined in the cultural import of Urhobo and Yoruba names, which are tailored as Indigenous education media that transmits Indigenous knowledge and ideologies that modify the people's worldview and define their identity.

Olivier and Fanon's views put forth the notion that decolonization has different facets and strategies. These facets are occasioned by strategies put in place by colonial powers retain colonial elite ideas and influences, management of transferred territories by Indigenous people, and self-reinforcing tendencies of each territory. In order to reinforce their political identity, most countries have also attempted to reinforce their cultural identities through a resurgence of Indigenous art and cultural configurations. Renaming of people, places, and things, as in the cases of Chinua Achebe and other African nationalists, and Ghana and other African countries that changed their English given names, were the first noticeable reordering program in most former colonies. The Urhobo and Yoruba cultural group were not left out of this decolonization strategy as it is commonplace for virtually every Urhobo or Yoruba child or person to be identified by an Indigenous name rather than an English name.

Decolonization and colonialism thus become expressed as binaries expressing conflicts of power. These conflicts can be ideological, intellectual, linguistic, social, or economic. This kind of mindset of inferiority is still prevalent in specific urban centers of Nigeria, where parents believe and contend that Indigenous names can be a barrier to their children being appreciated outside the shores of the nation. Such assumptions make apparent the idea that colonialism is far from being a finished business. This situation demands for, as Olivier and Fanon push, re-education of the people through systems of traditional knowledge and a complete reprogramming of the individuals in African and Nigeria into a new being that is unaffected by racial divide. This study attempts to be an avenue for such a reprogramming process by unraveling to its audience the beauty, aesthetics, and historical nature of the Urhobo and Yoruba names.

Ethnographic Survey of the Study Area

The Yoruba and the Uhrobo are two of the five largest ethnic groups in Nigeria. The Yoruba occupy the southwestern part of Nigeria, with an estimated population of 15.5 % of the Nigerian population. They are predominantly found in Lagos, Ogun, Oyo, Ondo, Ekiti, Kwara, parts of Kogi, and Edo states in Nigeria. They are also found in parts of the Republic of Benin and Togo. As a major ethnic group in Nigeria, much of its histories and heritage are captured in names imbued with meaning and reflecting Yoruba ancestry across geographical borders. While these names served as a cornerstone of Yoruba identity, colonialism's influence on Yoruba culture is evident in the introduction of foreign names, which presents a challenge to Indigenous established practices, potentially disrupting the connection between names and cultural heritage.

The Urhobo people on their part are a cultural group found in the South-Southern part of Nigeria. The Urhobo exist in the current Delta state in Nigeria with an approximate 1.5 million population, spanned across 24 kingdoms spread across ten Local Government Areas of the State. The people speak the Urhobo language, which defines their culture, its components and practices. The Urhobo operate a communal society where the Indigenous language plays a significant role in socialization, identity formation, creativity, and expression of beliefs, ideas, and worldviews. In order to actualize these, names play a significant role as people, places, beliefs, values, ideas, cultural creations, and everything must be given a name. Naming, among the Urhobo, is foregrounded by narrative and historical ethos. The circumstances, events, and experiences surrounding the birth of a person, place, or thing are often summarized philosophically, satirically, or paradoxically to name them. Hence, the names become a document of history and story, evoking narrative, meaning, and interpretation. Take the example of the name, *Ogbakworughworunmue*: The warrior that can defeat Orhughworun does not exist is a proverbial name that concisely documents the narrative of series of victories enjoyed by the Orhughworun community at various points in history. Such names abound among the people even though the influence of the English language and its attendant naming practices have taken its toll on the Indigenous culture, its languages, and practices.

Recently, there seems to be an awakening call from Urhobo scholars and societies such as the Urhobo Progress Union, Urhobo Studies Association, among others, for a revival of the use of the language in all family, communal, social, and other activities. Mainly, naming practices have gained focus since it is a convenient tool for telling stories and projecting cultural norms and values among the people. For instance, the 2005 publication by the Urhobo Historical Society, *Studies in Urhobo Culture*, contains a section dedicated to Urhobo names and naming practices. It interprets over 200 Urhobo names and conveniently serves as a mini-dictionary of Urhobo names, their spiritual, religious, and philosophical meanings.

The Impact of Colonialism on Indigenous African Names and Identity Crises

In Indigenous African society and cultures, a lot of factors are considered before giving a name to a child. There is a belief that a name conveys a lot about the origins, ancestry, future, and, in some cases, a person's destiny. The circumstances of birth and the impact such births have on the family are reflected in the newborn's given name. For example, among the Indigenous African people during the precolonial era, names were occasionally given to people based either on their striking resemblance with a family member or after the death of a family member, most especially that of a grandfather or grandmother, was recorded. Any child born after this occurrence is believed to be the reincarnation of the dead man or woman and named. There was no separating the African belief system from names. In Nigeria, among the Urhobo and the Yoruba ethnic groups, this has been the practice, until the contact between Nigeria, the Europeans and other civilizations significantly impacted the Nigerian languages and cultures, and this also reflected on names of people and places in the nation. This contact had implications on naming of people and places, which include change of names, misspelling or mispronunciation of names. The colonial masters used this method of replacing Indigenous Nigerian names with foreign or English names as a signal that they were entirely in charge of Nigeria's socio-political and economic affairs.

In one of the focus group discussions conducted among the Urhobo and the Yoruba youth age group, it was observed that in recent times, things have taken a new dimension regarding names and naming practices. The observation was that among the youths, they preferred English names and even devised a way of writing and pronouncing their names in an English style. As a result of this, it is obvious to hear foreign names such as Ivan, Rodney, Sean, Michel, and Diana among others among the youths. Some Indigenous names have also been changed to sound English. Nowadays, among the youth, it has become common to hear and see distorted Indigenous names made so purposely to modernize the name. Thus, you hear of Funky; originally Funke has now become Funk, while Akpofure is now written as Apos, to mention a few.[18]

Examining these trends among the youths, the distorting impact that the alteration of Indigenous names is bringing to the bearers' roots and the corrosion of the cultural values attached to those names is glaring. Inquisitively, these youths who preferred bearing trendy foreign names may have little or no understanding of the meanings of the foreign names they are bearing, yet they will rather bear those names as against their Indigenous names with rich cultural and historical meanings and which define who they are.[19] According to Odekunle, one of the respondents during the focus group discussion, borrowing from the words of Frantz Fanon, asserted that "the worst form of colonization is the colonization of the mind and the worst form of slavery is the slavery of the mind."[20] He further asserted that some Nigerian youths are victims of mental enslavement and mental colonialism. Identity crisis is the main reason people choose and prefer another culture to their own.

Dynamics of Decolonization, Encoding, and Expressing Naming Practices Among the Urhobo and the Yoruba

During the colonial and postcolonial eras, many Nigerians resisted the influence of colonialism on their culture, and in particular, this could be observed in the way and manner many began to drop their foreign or English names in favor of Indigenous names. One Nigerian nationalist and a former Deputy Mayor of Lagos in 1951, Mazi Mbonu Ojike, advocated a return to Indigenous Nigerian cultural values. He was recognized as the "boycott king" for his slogan "boycott the boycotters."[21] He schooled in America and upon his arrival back to Nigeria, he substituted his title Mr. for Maz, an Ibo title. He called for a boycott of anything foreign in favor of Indigenous Nigerian cultural values. He discouraged imitating the European way of life in his discourses, contending that Africans should content themselves with what they have. In demonstration of his belief in his advocacy of commitment to Indigenous cultural value systems, Ojike dropped his foreign name, jettisoned foreign cloths, and called for a rejection of the worship of any foreign God.

Another Nigerian nationalist who changed his name as a mark of resistance to colonial culture was Mojola Agbebi (1860–1917). After his parents were converted to Christianity, the father became an Anglican Catechist. When Mojola was born, he was named David Brown Vincent at baptism. He was educated by the Church Missionary Society (CMS). After his education, he joined the service of the Baptist Revival Service as a clergyman. While in the service of the Baptist church, he developed displeasure at the marginalization of Indigenous African clergymen. In response, he advocated the establishment of Indigenous African churches, which he pioneered in 1888 with the establishment of the Native Baptist Church in Lagos. This became the first Indigenous church in West Africa. Along with this came the change of his baptismal name David Vincent Brown to an Indigenous Yoruba name Mojola Agbebi in 1894.[22] All these were the

Table 5.1 Urhobo names for people.

S/N	*Names*	*Meaning*	*Narrative and historical content*	*Inherent practice, belief, and values*
1	Aruoture	We have now attained greatness	A tale of attainment of socially expected marks of achievement and a history of time-marked age-grade	Hard work, resilience, doggedness, and healthy competition.
2	Omotejohwo	A girl is also a human being	A tale of recurring birth of female children and resultant marginalization	Respect for humans rather than gender.
3	Ovwokpaegajitorho	The baby strap of the barren eventually hung on the line	A tale of delay in childbearing and maternal stratification in a large family setting and the eventual birth of a child with a sense of fulfillment	Hope, patience, and respect for those waiting to achieve a "feat"
4	Omokughebe	The child is a mirror	The story of a child with exceptional and unexplainable beauty is considered reflective beauty, hence the mirror metaphor	Creativity: the ability to draw comparisons
5	Ejuvwierhe	Let the kingdom come	The name tells the story of one waiting patiently for things to get better in the family. It conveys the narrative of hardship relieved by the hope of a better future	Hope, contentment
6	Ejuvwehwe	I must not be mocked	A tale of one who is not willing to stop trying for success due to the fear of being ridiculed	Pride and determination
7	Emusherearhe	Let food fall from the sky for us to eat	A paradoxical rebuke for people who would not work	Hard work
8	Ohwovoriole	The one that eats yam alone	A satirical mockery of persons that exhibit greed and selfish behavior	Generousity, altruistic
9	Omonigho	A child is of more worth than money	A tale of being in a situation of having to choose between one's child and wealth or a source of livelihood	Commitment and ability to prioritize in decision-making

(*Continued*)

Table 5.1 (Continued)

S/N	*Names*	*Meaning*	*Narrative and historical content*	*Inherent practice, belief, and values*
10	Emoyigho	Children are financial wealth	Investing one's resources in one's children, considering the children's success as one ultimate wealth	Value for the right things and consideration for the future
11	Omorhu/ Omoavworho	Children are one's greatness	A tale of how society values children as a mark of greatness. So, a man with wealth but without children lacks greatness	Perspective to life
12	Erhuvetine	I am being oppressed here	It tells a tale of a person being covertly oppressed and lacking an avenue to cry for justice due to cultural expectations of not getting divorced or fighting back	Expression
13	Ughumiovakpo	It is death that spoils life	A philosophical tale on the pains and effects of death on life and relationships	Accepting that which cannot be changed
14	Akponehwe (Erivwinejowo)	The living demanded his death (the ancestor acquitted him)	An ideological tale that unveils the worldview of the Urhobo on the dialectical relationship between the two spheres of their world: the spirit realm and the realm of living	Belief in the supernatural (gods and ancestors: the living dead)
15	Ighofose	Money destroys friendship or marriage	A tale of "modern" women who are not ready to continue with any love or marital relationship once financial value ceases to exist for them	Selflessness, commitment, sacrifice

Table 5.2 Yoruba names for people.

S/N	*Names*	*Meaning*	*Narrative and historical content*	*Inherent practice, belief, and values*
1	Kasimawo	Let's still be looking at him	A name given to the only surviving child	
2	Ige	This is a natural name the child came with from birth	A name given to a child naturally from birth because the child came out of the womb through his/her legs and not the head	Such a child is regarded as a unique child
3	Tokunbo	He came from the sea	A name given to a child born abroad	
4	Ojedele	This name typifies the linage of traditional worshipers	A child born into the lineage or family of Masquerade worshipers	
5	Adebambo	The crown came with me	A child born into a royal lineage.	
6.	Ifabamidele	God has followed me home	A child born into the traditional worshippers of Ifa	It is common among traditional Ifa worshippers
7.	Akintola	A valiant man with wealth	A name given to a child in expectation that that child will be bold and courageous in life	
8.	Ogunniyi	God of Iron has honor	A name given to a child born into the family who worships the god of iron worshippers	A traditional religious name
9.	Folasade	A child living in abundance of wealth		

various methods Nigerians used in shedding off the colonial insignia and a mark of decolonization and a return to Indigenous African culture and values system.

The dynamics of decolonizing names and culture were not only peculiar to Nigeria. In the former Congo, the then military head of state Colonel Joseph Desire Mobutu, having come to power, came up with a policy known as the Authenticity Campaign, which he believed would guide Congo back to its cultural past. He replaced all places in Congo bearing foreign names with Indigenous names; he changed Congo the name of the country to Zaire, changed his name to Mobutu "Sese Seko," and mandated all Congolese bearing foreign names to do likewise.

All these are glaring facts that Indigenous African people need to go back to their roots and revive the traditional rich African culture through the decolonization of foreign names among the African people. This will help to preserve the African Indigenous names and cultural heritage.

Conclusion

From this study of decolonizing names and dynamics of cross-cultural naming practices among the Urhobo and the Yoruba ethnic groups in Nigeria, names have historical meaning and are the sum of the reflection of the culture and life knowledge of the people. From this study, names and naming practices among the Urhobo and the Yoruba are sources of identification marks. Names that people bear typify a symbolic value as a repository of history about individuals and their societies. From the study areas, it is noted that names are value-laden due to the historical narrative behind those names and how the Urhobo and the Yoruba conducted their lives both in the past and present.

In the study area, sometimes names given to children are grounded on their cultural beliefs and connected with specific cult groups or deities worshipped by the family. Names and naming practices among the Urhobo and the Yoruba are generally an emotional, mental, and linguistic affair motivated by diligent soul searching. The studies discovered that names are relevant and considered an essential social activity of the two ethnic groups, and as a result, names are carefully selected.

Furthermore, the study reveals that the clamour for decolonizing names among colonized people, as advocated in this study, is in tandem with the spirit and doctrine of the African renaissance currently pervading the entire continent. The advantages of decolonizing names among the Urhobo and the Yoruba ethnic groups in Nigeria are enormous. The bearing of foreign names or misspellings of Indigenous names is a symbol of colonialism, which erodes the rich culture of the Indigenous people and their heritage. The Indigenous African names and culture can still be preserved and must be preserved through the people. And this can be done by reverting to the Indigenous names and naming practices. Decolonizing names and naming practices will propel the preservation of Indigenous names and culture, as well as the Indigenous languages and traditional cultural values of the people.

Notes

1 Punch, "Advertise Your Change of Name in Saturday Punch, Nigeria's Most Widely Read Newspaper," *Lagos: The Punch Publishers*, February 2, 2022.
2 Orhioghene B. Akpomuvie and May O Esiri, "A Cross Cultural Study of Names and the Unearthening of Naming Practices in the Three Ethnic Groups in Nigeria," *PalArch's Journal of Archaeology of Egypt/Egyptology* (2021), pp. 3680–3691, 3681.

3 Gbenga Fasiku, "Yoruba Proverb, Names and National Consciousness," *The Journal of Pan African Studies* 1.4 (2006), p. 12.
4 Uluocha O. Nna, "Decolonizing Place Names: Strategic Imperative for Preserving Indigenous Cartography in Post-Colonial Africa," *African Journal of History and Culture* 7.9 (2015), pp. 180–192.
5 A. Odesanya, S. Oloruntola, and A. Kunle, "Names as Message Vectors in Communication: Ideological Analysis of Traditional Yoruba Personal Names from Ifa," *Journal of the Linguistic Association of Nigeria* 20.1 (2017), pp. 248–259.
6 Cheng Patrick Low Kim, "The Confucian Rectification of Names and People/Human Resources Management," *Educational Research* 3.8 (2012), pp. 662–668; M. Heiko, "Corpus Linguistic Onomastics: A Plea for a Corpus-Based Investigation of Names," *Journal of Onomastics* 68.2 (2020), pp. 88–103; I. Mackenzie, "What's in a Name? Teaching Linguistics Using Onomastic Data," *Language* 94.4 (2018), pp. 293–310; A. Koopman, "The Post-Colonial Identity of Durban," *Oslo Studies in Language* 4.2 (2012), pp. 133–159.
7 University of Minnesota Libraries, "The Element of Culture in Sociology: Understanding and Changing the Social World," *University of Minnesota Libraries*, 2010. Retrieved November 2, 2022, https://0pen/lib.umn.edu/sociology/part/chapter3-culture.
8 M. J. Malukeke, "Culture, Tradition, Custom, Law, and Gender Equality," *PER/PEL Online Journal* (2012), p. 2.
9 Malukeke, "Culture, Tradition, Custom, Law, and Gender Equality," p. 2.
10 J. A. Banks and C. A. Banks, *Multicultural Education: Issues and Perspectives* (New York: Hoboken, 2016), p. 27.
11 C. Kluckhohn and H. W. Kelly, "The Concept of Culture," in *The Science of Man in the World of Culture*, ed. R. Linton (New York: Columbia University Press 1945), pp. 78–105, 86.
12 Campbell M. Segall and M. Herskovits, *The Influence of Culture on Visual Perception* (Indianapolis, IN: Bobbs-Merril Co., 1966), p. 9.
13 UN, "UN System Task Team on the Post 2015 UN Development," 2015, https:/www.un.org/milleniumgoals/pdf/Think%20Pieces/2_Culture.pdf, p. 2.
14 O. Nnoli, *Ethnic Politics in Nigeria* (Enugu: Fourth Dimension Publishers, 1978), p. 35.
15 Eze Dons, "Nigeria and the Crisis of Cultural Identity in the Era of Globalization," *Journal of African Studies and Development* 6.8 (2014), pp. 140–147.
16 B. Olivier, "Decolonization, Identity, Neo-Colonialism, and Power," *Phronimon* 20 (2019), pp. 1–18, 1–2.
17 Franz Fanon, *The Wretched of the Earth* (New York: Groove Weidenfeld, 1963), pp. 35–36.
18 Ayomide, Olumide, D, interview by Abayomi S. Israel. *Mr.* (October 20, 2022).
19 Foluke, Ajayi, A., interview by Abayomi S. Israel. *Mrs* (October 20, 2022).
20 Odekunle, Adebayo, interview by Abayom, S. Israel and i. *Mr.* (October 20, 2022).
21 Gloria Chuku, "Mbonu Ojike: An African Nationalist and Pan-Africanist," in *The Igbo Intellectual Tradition: Creative Conflict in African and African Diasporic Thought*, ed. Gloria Chuku (London: Palgrave Macmillan., 2013), pp. 89–117.
22 Rina Okonkwo, "Mojola Agbebi Apostle of the African Personality," *Présence africaine* 114 (1980), pp. 144–159.

6 Translanguaging Names, Decolonizing Language

What Makes a Name "Chinese"?

Federica Guccini

"This girl with that really Dutch name is coming. . . . You'll be able to see what a real Dutch person looks like." This is how Maia's Dutch professor announced that Maia would be joining Dutch lessons soon, according to Maia's classmates. When Maia did join, the professor asked if she was in the right class. In our interview, Maia told me, laughing, "When I showed up being Asian . . . , he was really so surprised."[1] Maia was born in China and adopted into a Dutch American family as an infant. Despite Maia's upbringing and her "really Dutch name," the instructor's expectation that a "real Dutch person" would join his course seemed compromised by Maia's appearance. He likely operated on a popular understanding of names as tied to a particular language or language community, which in turn are attached to expectations regarding name bearers' ethnoracial identities.

What would it look like if we challenged monolingual and monocultural classifications of names? How can we decolonize our conceptualization of names and naming practices? These questions guide my theoretical approach, in which I suggest that naming practices and their social meanings are entwined in colonial ideologies of language and race. Such raciolinguistic ideologies—beliefs and judgments about the perceived correlation of language and race—are rooted in colonial thinking "of superiority and inferiority that turns differences between people, knowledge, expressivities, or languages into values."[2] It is, therefore, crucial to understand that our categorization and hierarchization of languages are colonial ideologies in and of themselves. I argue that we should also apply this lens to names and naming practices.

Personal names are commonly understood to be mere referents for people; however, naming also invariably carries social and political importance.[3] Language ideologies that position a personal name both as a word and a symbol for a person can, therefore, not be detached from people's social relationships and identities.[4] These considerations explain why paying attention to colonial ideology in naming practices is so crucial. In this vein, names are often approached from the standpoint of monolingual bias, which sees names as belonging to only one language or speech community. This view persists even though many names

DOI: 10.4324/9781003414773-7

have, over time and across the globe, gone from being "foreign" names to staples in local naming culture. We thus need a framework that allows us to talk about names and names research without reinforcing this monolingual bias.

I propose to explore naming practices through the lens of what I call Translanguaging Names Theory. Ofelia García and Li Wei define translanguaging as "speakers' construction and use of original and complex interrelated discursive practices that cannot be easily assigned to one or another traditional definition of a language."[5] I use the term here to provide a new way of thinking about the relationship between names and language(s): Instead of being conceptualized as bound to specific languages or speech communities, Translanguaging Names Theory positions names as embedded in unitary semiotic repertoires that more accurately reflect name diversity and lived experiences of carrying names that do not fit colonial conceptualizations of language boundaries.

I follow Gabriela Veronelli in framing coloniality as "a global process [which] has local manifestations that depend on particular colonial histories, geographies, cultures, forms of colonization and racialization, and negotiations with and resistances to imperial power."[6] While Translanguaging Names Theory should apply to naming practices everywhere, my conceptualization draws specifically on observations from two case studies. Case Study A is based on archival research and interview data collected between 2018 and 2021 regarding identity and language practices in the Hakka Chinese community in the Indian Ocean island state of Mauritius.[7] The findings show how colonial linguistic processes and ideologies shaped new naming patterns among colonized populations. Case Study B focuses on interviews with Chinese students at a midwestern college in the United States and a university in central Germany, conducted from 2015 to 2016, as well as online research on public discourse regarding Chinese people's name choices.[8] This second case study illustrates why Translanguaging Names Theory is crucial in the endeavor to decolonize how we think about names and naming practices.

Ofelia García et al. make it a point to emphasize the *people* in their work rather than the *languages* which they speak. Similarly, an integral part of a decolonial perspective on naming practices should be name bearers and their lived experiences rather than the names themselves, lest we dehumanize what it means to live and navigate the world with one's name(s).[9] After all, "all people deserve to be called correctly and identified by the name of their choice."[10] However, dominant languages and their speakers hold power even in multilingual or multicultural social spaces. This is reflected in common expectations that people with non-dominant language backgrounds shorten or change their names or even choose a new name entirely to accommodate others. Subverting these expectations can be an exhausting and thankless endeavor for name-bearers. Shifting how we think about the intricate connections between names, languages, and cultures is a first step in the right direction. In this sense, ethical and politically informed research should consider that "the study of language can never be *just* about language."[11] Therefore, Translanguaging Names Theory must incorporate a more practical element in the form of actionable guidelines and methodological tools for centering lived experiences of names, naming, and their social implications.

A Decolonial Lens for Naming Practices: Toward a Translanguaging Names Theory

Anthropology has long struggled with perspectives of decolonization, given that the very "social and historical origin of many of our key concepts is the colonial situation."[12] Decolonization thus cannot mean a return to an inexistent precolonial anthropology; instead, there is much need for an "anthropology of liberation"[13] with a practical dedication to interdisciplinarity in theory and methodology.[14] In this sense, decoloniality is not the mere "absence of coloniality, but rather the serpentine movement toward possibilities of other modes of being, thinking, knowing, sensing, and living; that is, an otherwise in plural."[15] To this end, I situate this work not solely in (linguistic) anthropology but also in sociolinguistics, onomastics, and other related fields, with the aim of finding a potential names theory that takes into account the coloniality of naming but moves beyond it by thinking of names in an "otherwise" way.

Decoloniality-oriented linguists and anthropologists have done substantial work to dismantle the colonial theory framework surrounding our understanding of language(s) and linguistic practices. In the era of colonial linguistics, the act of writing grammars and dictionaries of Indigenous languages in the colonies was not merely a research project so much as it was entangled in harmful colonial politics. Indigenous languages were instrumentalized "to impose a Eurocentric epistemology on Indigenous populations . . . as part of colonial domination."[16] Languages and language boundaries are constructed, categorized, and recognized according to sociopolitical developments in a world that continues to uphold the colonial system.[17] As such, when we discuss the coloniality of language, we also discuss the "*coloniality of power* in its linguistic form: a process of dehumanization through racialization at the level of communication."[18] Veronelli here draws on the concept of "coloniality of power" coined by Aníbal Quijano, who identifies as the two pillars of the concept (a) the focus on production under world capitalism and (b) the naturalization of "race" as a supposedly biological factor for social stratification.[19]

The dehumanizing process of racialization as a form of coloniality of power is also at the core of the field of raciolinguistics, which critically engages with ideologies that see language and race as inherently and naturally connected.[20] In their manifesto "Rejecting Abyssal Thinking in the Language and Education of Racialized Bilinguals," García et al. make the case that perceived boundaries between named languages need to be challenged theoretically because they are, in practice, already challenged by the multifaceted speech of the racialized bilinguals at the center of their research. They argue that this speech can be described as *translanguaging*, which denotes the use of a unitary semiotic repertoire rather than multiple repertoires made up of various languages.[21] Wei Li has also described this as "both going between different linguistic structures and systems . . . and going beyond them."[22] Suresh Canagarajah similarly sees the labeling of separate languages as "an ideological act of demarcating certain codes in relation to certain identities and interests" in stark contrast with a more commonplace "translingual practice."[23] Translanguaging can thus be seen as a critical response to the coloniality of power and colonial language ideologies.

This rich framework of decolonial theory on language, race, and power forms the basis of my Translanguaging Names Theory. Names are rarely exclusively used in a singular linguistic context. Names being spoken and written across languages and language barriers often results in pronunciation shifts through the use of different accents or mispronunciation (accidental or deliberate), official name changes for legal or social reasons, and alternate naming practices such as nicknaming or the adoption of a new name. Transgressions such as "renaming, denaming, and misnaming of [people] from linguistically marginalized and ethnoracially minoritized backgrounds" happen all too commonly as well.[24] As I have shown in the opening example of Maia and her Dutch professor, names carry expectations regarding people's ethnoracial identities and can function as markers of "presumed linguistic ability."[25] These assumptions stem from monolingual bias: the common misconception framing "the prototypical human as having only one language" with which they are inherently linked.[26]

In reality, multilingualism has always existed, and finding a language without any external influences is impossible.[27] The same applies to names: Naming practices do not exist in linguistic or cultural isolation and have always influenced each other (see, for instance, Möller 2019 on multilingual place naming in southern Africa). While this may be an inevitable result of language contact or linguistic proximity, names can also be translanguaged intentionally. The latter is evident, for instance, in García and Li's discussion of a sign that reads "I ☘ O'Bama," interpreted either as "I love Obama" or "Ireland loves Obama."[28] Barrack Obama's name is here spelled in the Irish style with an apostrophe following the initial O. Intentional name translanguaging will not always be as overt as this example, which is clearly meant for humorous purposes. Still, it is important to note that people might translanguage names for specific reasons.

Moreover, there is "property-like potential in names to transact social value."[29] Such commodification of names works on various levels. It can be understood as a phenomenon of the global linguistic market, in which names function as strategically accumulated linguistic capital.[30] On another level of commodification, many may understand names as the intellectual property of a (monolingual) speech community. In this vein, names can even become products to be purchased, with naming practices turned into an authenticated and marketable service, as Case Study B will illustrate. A translanguaging approach to names is thus also useful in the critique of production-oriented world capitalism that is emblematic of coloniality.[31]

Case Study A "Colonized" Names Among Chinese Mauritians

Mauritius is a small island state in the western Indian Ocean with about 1.3 million citizens. Over almost 400 years of colonization under Dutch, French, and British rule, colonizers enslaved and indentured populations from West and East Africa, India, and China to work on the island's sugar plantations. Postcolonial, independent Mauritius proudly labels itself a multicultural "rainbow nation," in which Chinese Mauritians—also known as Sino-Mauritians—account for about 3% of the population, according to the last available census data from the 1980s. The majority of Sino-Mauritians are Hakka, an ethnic sub-group of the Chinese Han majority, whereas a minority of about 10% are Cantonese. Hakka and Cantonese are not spoken as much in Mauritius anymore due to language shifts to the *lingua franca* Mauritian Creole (*Kreol morisien*), French, and English, as well as the rise of Mandarin learning in Chinese diasporas.[32] However, the Mauritian Chinese family name is a strong reminder of Hakka and Cantonese language presence in the community.

Mauritian names have not been studied much, but anthropologist Thomas Eriksen mentions them in a prelude to his discussion of linguistic and cultural diversity in Mauritius:

> Initially, it may seem a stupendous task to try to make sense of social categories and cultural patterns in a society that is simultaneously a member of *La Francophonie* and of The New Commonwealth; where fifteen languages may be spoken, according to official statistics; where the currency is the rupee, whose name is usually pronounced in a vaguely French-sounding language (but the banknotes have text in English, Hindi and Tamil); where people can possess names like Françoise Yaw Tang Mootoosamy; and where the second-in-charge of the Catholic Church turns out to be a man named Amedée Nagapen. The pages that follow . . . are an attempt to make sense of this apparent chaos.[33]

The "apparent chaos" Eriksen identifies here is defined by the various linguistic contexts that co-exist in Mauritius, resulting in translingual speech, writing, pronunciation, and naming. Eriksen seems to mark the names Françoise Yaw Tang Mootoosamy and Amedée Nagapen as surprising or confusing without explicitly specifying any context for either such a reaction or the names themselves. The premise therefore exhibits an unspoken element of monolingual bias.

Family names in the Mauritian Chinese community are longer than most family names in mainland China. Sino-Mauritian scholar Clement Chan speaks of "colonised surnames" when referring to the typical Chinese family name structure in Mauritius, typically consisting of two or three syllables.[34] This is a significant departure from Chinese naming practices in China, where individuals usually carry a one-syllable family name.[35] Sino-Mauritian family names are also spelled in Romanized fashion, which marks another important difference. Consider the following example: Chinese indentured laborers and other immigrants arriving in Mauritius would have had names such as Chu Wei Chuen, in which Chu would be the family name, followed by the given name Wei Chuen, of which one syllable would be a generation name shared by all male brothers and cousins of the same generation, and the other would be the individual name not shared by anyone else.

Chu Wei Chuen was a Hakka Mauritian from Meixian, Guangdong Province, who landed in Mauritius as a teenager in 1887.[36] In Chinese script, his name would have been spelled 朱维勳 (*Zhū Wéixūn* in Mandarin, *Zu*^{1} *Vi*^{2}*hiun*^{1} in Hakka). Migrants seeking passage out of China would typically go through Hong Kong, where they would board a ship to Mauritius. British colonial officers there would write down the names of ship passengers. Hong Kong being a Cantonese-speaking area, British officers would have lacked familiarity with Hakka pronunciation, and thus noted the name in Cantonese style. Chu Wei Chuen therefore bears little resemblance to the original Hakka pronunciation.

One of Chu Wei Chuen's sons became a well-known politician and businessman in Mauritius. His face adorns the Mauritian 25 Rupees bill (see Figure 6.1). His name was Moilin Jean Ah-Chuen. Note that 朱, or Chu, is not the family name here. In Mauritius, the family name of Chu Wei Chuen's descendants became Ah-Chuen. Chuen is the last syllable of Chu Wei Chuen's name, part of his given name. Combining it with "Ah" makes it a typical form of a Chinese nickname and may thus have been what others called Chu Wei Chuen. Instead of Chu, Moilin's family name was his father's nickname.

The nickname [Ah] + [given name syllable] form is not the only type of family name in the Sino-Mauritian community. More commonly, Sino-Mauritians

Figure 6.1 Mauritian 25 Rupees bill depicting Sir Moilin Jean Ah-Chuen. Tonton Bernardo, CC BY-SA 3.0 <https://creativecommons.org/licenses/by-sa/3.0>, via Wikimedia Commons.

have a three-syllable family name. Study participant Roland Tsang Kwai Kew has one such name. The first in his lineage to arrive in Mauritius was a man named Tsang Kwai Kew. His family name was Tsang, and his given name was Kwai Kew; now, his descendants carry his full personal name as their family name. In an interview with Cécile Leung, she described how three-part family names came into existence:

> To register the birth of a child, they ask you, "What is the name?" Okay. And the father give the whole name [of the child]. . . . "What is the father's name?" They wanted to say, "the surname." . . . And [now] I have on my back twice my family [clan] name, my own [Chinese] given name, and my father's given name! Everybody is like that in Mauritius. (Interview with Cécile Leung, 2019)

The miscommunication of the father's full name as the family name thus created a unique local family naming practice. Instead of merely passing the clan name down the patrilineage, the Sino-Mauritian family name is more of a patronym. Observing how the Sino-Mauritian family name underwent two significant changes—firstly, the transliteration into Romanized, Cantonese-style spelling in Hong Kong and, secondly, the misregistration of family names at birth in the colony—thus prompted Clement Chan to speak of "colonized" surnames. After all, these names do not (exclusively) follow Chinese naming practices but new norms that emerged out of the local colonial context in Mauritius.

Of course, paired with the Christian given names of most Sino-Mauritians (Roland and Cécile, for instance), other naming practices from the French and British colonial influence enter the scene. While some interviewees did not have a Christian name or used their Chinese names exclusively, the majority of Sino-Mauritians with whom I spoke went by a Christian name in English or French pronunciation. The combination of naming practices and the length of the full name was often bemoaned as troublesome in official and administrative matters, such as air travel, identification documents, or school registration. Authorities and officials would often not immediately recognize the name as Chinese—not even Chinese speakers, who frowned at the long family names. This raises the question: What makes a name "Chinese" or "English" or "French"?

Case Study B Translanguaging Naming Practices Among Chinese Students

In Case Study B, I interviewed Chinese students at a college in the American Midwest and a university in central Germany about their names. Their name repertoires often included not only their Chinese given names but also alternate names for social encounters with non-Chinese people, for instance, while studying or working abroad.[37] Such names have been called "English" or "western" in public discourse. Consider these newspaper and blog headlines:

(a) *The Atlantic*: Hong Kong Loves Weird English Names
(b) *Public Radio International*: "Hello, My Name Is Yes," and Other English Names in China
(c) *The World of Chinese*: Hilarious English Names Chinese People Choose
(d) *eChinacities*: Chinese People's English Names—Is There a Method to the Madness?
(e) *The Nanfang*: Laowai [meaning "foreign(er)"] Entrepreneur Wants to Rid China of English "Stripper Names"
(f) *The Telegraph*: US entrepreneur vows to rid China of "farcical" Western names: Lindsay Jernigan hopes to liberate young Chinese from bizarre adopted names including Lady Gaga, Twinkle, Pussy, Elvis—and Washing Liquid[38]

These headlines exhibit judgment of Chinese people's naming choices, which ranges from amusement and declarations of madness to charity work and liberation by way of a "white savior" putting an end to the practice. Names that are deemed funny, inept, inappropriate, or otherwise unusual are listed and analyzed, as well as belittled. Names of this category mentioned in the articles include popular examples such as Seven and Eleven (often used together for a joke about the convenience store), Washing Liquid, Watermelon, and Cinderella. However, Sapphire, Candy, and Cherry also make an appearance, as part of the "stripper name" categorization, despite their common use in English naming culture.

The power structures behind the hierarchization of languages and names in these articles need to be considered: English holds a different position than Chinese in the global economy, and so do names deemed "English" or at least compatible with English. As Pierre Bourdieu states, "linguistic relations of power [are] based on the unequal distribution of linguistic capital."[39] This puts into context the launch of a (no longer active) website called BestEnglishName.com, created by Lindsay Jernigan, a Shanghai-based American, who is in fact the entrepreneur from the headline (f) above, who declared to "rid China of 'farcical' western names." In her introductory video on the website's homepage, Jernigan explains the need for this website and how it works:

> I got the idea for this website from working with some colleagues that had strange English names, talking to my western friends who were always joking about their colleagues and people they had met with strange English names. I realized that there might not be a very good resource available for Chinese people looking for an English name, so I decided to create this website for people like you looking for an English name. At BestEnglishName.com, you can read about hundreds of English names for free. You can also read about western naming culture, learn more about how we came up with these names, [and] why we use them [. . .]. You can take a quiz that will automatically generate five English names—just for you![40]

The company used the discourse of "weird" names to create a business dedicated to eliminating ridicule that Chinese people might face with a "strange" English name. They claimed to provide them with "better" options, if not the "best," which commodified the name as a marketable product. A sense of "authenticity" was produced alongside these names due to their apparent connection to native speakers, who hold an inherently "legitimate" position in the global capitalist economy to provide such services. If names can be "bought" like this, transferred from one economic party to another, they can be perceived as "property"—but whom do they belong to?

The company suggested that the website was the single best resource a Chinese person could use to find an English name because it was "China's only resource for English names created by a westerner and native English speaker."[41] This reveals a monolingual bias that binds a name and its valorization to one speech community only, turning the name into the community's intellectual property. In this light, English names belong to English speakers and are theirs to be given away. This effectively delegitimizes Chinese resources for "English" names and denies Chinese people's ability and agency to choose their own names. On the other hand, no matter the linguistic competencies of monolingual English speakers, "unusual" name choices among them are rarely treated as a lack of appropriate knowledge of English but rather unique choices or individual mishaps. Contrasted this way, name commodification is tied to raciolinguistic ideologies that treat racialized bilingual speakers (of English) as deficient.[42]

What is often overlooked in the discussions is the way Chinese "English" names are not just borrowed from English but Sinicized in the manifold ways that they are created and used. Translanguaging is present in many of these alternate naming practices. "Frankly, a name isn't just a name, it could bring good or bad affection to you. Choosing an English name can be very serious," explains a research participant in Paul McPherron's study.[43] This quote contains clues to how Chinese people do not just start using alternate names at random but carefully choose and craft their internationally usable aliases following Chinese naming practices: Chinese given names are often selected with auspicious and astrological factors in mind, and such considerations for a fortunate name may influence Chinese name choices across language contexts.[44] As such, a sense of Chineseness may be retained even in names that others do not immediately read as Chinese. To illustrate: Chinese given names are typically chosen according to words' literal meanings, often related to a person's character or birth circumstances. When Xueqing, one of my interviewees, was born, the sun came out after a snowfall, and her parents named her "snow" and "sunshine" accordingly. Transferred to translingual naming practices, this leads to many Chinese people choosing names with literal meanings even when this is not necessarily typical in the naming culture in which they use those names. In my study, for instance, some participants were named or knew people who used the names Jade, Crystal, Sky, Shadow,

Key, or Timekeeper (TK for short). Yunan used the nickname Soda for her social media because the first character of the Mandarin word for "soda" matched her family name. However, she purposefully did not introduce herself as Soda to people abroad because she anticipated that they would find it strange.

The abundance of Chinese name options (because every word could also become a name) sometimes results in a preference for "unique" names over "common" names. Many of my interviewees were aware of approximately how many people they shared their names with and would often comment on name frequency. The tendency to opt for unique names tends to carry over to translingual name choices as well: Yuanwei told me that when she chose Coco as her "English" name, she thought it would be quite unique but found out that it is a very popular name among her peers (in part due to Coco Chanel's fame). Coco kept her name and accepted its lack of uniqueness, but others may avoid this outcome by using a slightly different spelling, such as Logan turning into Login or Megan into Megin.[45]

One interviewee from Hong Kong, where English names are common due to the British colonization history, told me that she and her sister were named Viv*ian* and Lil*ian* because their parents wanted an equivalent to the Chinese generation name in the shared syllable *-ian*. In Case Study A, there were also instances where Mauritian Chinese participants had to choose a Christian name at baptism. One research participant, Hok Fen, chose Stephen as his baptism name, pronouncing it in a more French way so that *-phen* would resemble the pronunciation of Fen. Case Study B participant Huimin told me that in mainland China, many parents would choose names or nicknames for their children that could be easily Anglicized, like Lili, which means "beautiful" in Mandarin but could easily turn into the name Lily. Names with repeated syllables, like Lily, Lulu, and Coco, for instance, follow a typical Chinese nickname pattern, making them very popular among Chinese people seeking an internationally usable name. Some Chinese websites for "English" name choices also advertise names that are *haoting* ("good-sounding"), which is a highly valued quality in Chinese naming practices. On the other hand, some names can carry pejorative meanings, such as Charlotte, which sounds similar to "kill her" in Mandarin.[46]

These naming practices and choices suggest that monolingual bias and colonial ideologies of languages as separate entities need re-framing. There are Chinese naming elements present even when the name is thought to be "English" on the surface level. Translanguaging practices challenge the idea of names as bound to one language or speech community, and they expose the ignorance behind the ridicule that racialized and non-native speakers of English may face when using a broader semiotic repertoire to choose a name for themselves or others.

Discussion and Implications: Are All Names Translingual?

The case studies I use to support my argument focus on the names that are, in one way or another, related to Chinese naming practices. I make the case that deviations from anything strictly seen as a "Chinese" name is not to be categorized as "non-Chinese"—or another language such as "English," "French," or 'Dutch"—but rather as a form of translanguaging. Multiple language contexts shape these names, and the described naming practices transcend singular language contexts into what Li Wei calls "translanguaging space."[47] Of course, given that my observations are based on two particular case studies, the question remains whether Translanguaging Names Theory also applies to other contexts. In other words: Are all names translingual?

Frustratingly, the answer cannot be a simple yes or no, especially with a decolonial perspective in mind. I believe that we should avoid *strict* categorization of names by language or speech community because all names are *potentially* translingual, and translanguaging is certainly common in naming practices. To say otherwise would reinforce colonial ideologies of raciolinguistic boundedness (race and language as natural, fixed entities) and monolingual bias. However, I want to be mindful of the possibility that such a claim could be detrimental to the very decolonial movements it is meant to support. Mary Bucholtz describes the phenomenon of "indexical bleaching" in names that "allows the literal reshaping of ethnoracially marked names—phonologically, orthographically, and even lexically—in ways that reduce their ethnoracial specificity."[48] This, too, is rooted in colonial politics and ideologies, as Black and Indigenous people were forced to discard their names as part of linguistic colonial oppression, and immigrants are frequently renamed and ridiculed because their names are deemed "difficult."[49] To mark all names as translingual could thus dismiss the tremendous pain of being stripped of their culturally specific names. I am also not suggesting that people should be able to freely appropriate names with certain racial, cultural, or religious connotations. After all, the coloniality of names is rooted in the "coloniality of power"; hence, name choices are linked to social status, power, and privilege.[50]

What I am suggesting is that names cannot be sorted into strict linguistic boundaries. The unboundedness of names is evident in the popularization of "foreign" names in language contexts in which they did not first emerge. Many common names in Anglo-American and European contexts are Greek (i.e.,

Alexander), Latin (i.e., Lucas), or Hebrew (i.e., Matthew) in origin, indicating a long-established history of translanguaging in naming practices. The "English" name Kevin was popularized in countries like France and Germany after the widespread success of the 1990 movie *Home Alone*; however, the name is originally Irish (Caoimhin).[51] The examples of Case Study A show how colonialism introduced new elements into established naming practices, and Case Study B further illustrates how translanguaging is used to creatively and agentively bridge linguistic contexts to make a name recognizable in various translingual encounters.

My own name is another example. When people see or hear my name, most assume that I am Italian (or Spanish). When they meet me, they may be confused to learn that my accent and/or language use do not match their expectations. In fact, I am half-Italian and half-German, which is ironic given that the "Italian" name Federica originates from the Germanic words *frid* and *ric* for "peace" and "ruler." I pronounce the "r" in "Federica" in (at least) three different ways depending on the language context. Italians say "Federica" with a rolled [r], in Germany, I was "Federica" with a uvular [ʀ], and since moving to Canada, I have—somewhat reluctantly—started pronouncing it as "Federica" with [ɹ] for easier recognition in an English-speaking context. I struggle to respond to questions about the proper pronunciation of my name because, to me, there is not one right way to pronounce it. I expect that other Federicas around the world may disagree with me, as being namesakes does not mean that we share identical experiences.

The divergence of experiences around names and naming, and lived experiences of name bearers more broadly, should be centered in anthroponomastics—which is, after all, the study of human beings' personal names. In a call for enriching theory with accounts of lived experience, Athur K. Spears finds that lived experiences can both "challenge and amplify observation-based research."[52] The study of personal names, of course, offers a wide range of opportunities to engage with lived experience in the research process. Most people have something to say about their name, and beyond that, what it is like to *live with* their name because "names are always implicated in social relations."[53] Qualitative methodologies can highlight lived experiences by letting people tell their stories, for instance through ethnographic research, interviews, open-ended survey questions, experiential models such as written, visual, or otherwise performed contributions from participants, and observations of everyday name encounters. Most importantly in the context of Translanguaging Names Theory, accounts of lived experiences can tell us more about people's actual practices and uses of names across perceived linguistic boundaries than studying the names themselves can. After all, behind names are the people who carry them.

I call here for an approach to names that reflects and respects the fact that names are not stuck in geographic or linguistic space. To illustrate precisely why we should think of names as unbounded from monolingual contexts, let me cite a counterexample: In a sociological study on immigrant names in Germany, Jürgen Gerhards and Julia Tuppat assert that "immigrants who are better integrated into the host society are more likely to choose names from that host country."[54] They cite linguistic and religious proximity as factors for similar name pools between two countries, using Italian and German names such as Emilia and Marco as an example. This is an oversimplification of the reasoning behind name choices, dismissing factors such as job security, cultural safety, multilingualism, or personal preference. Elsewhere, I have stated that such ideologies of names being either common or uncommon—even a marker of integration or acculturation versus "ethnic closure"—in a particular language or country "places them in a binary opposition that has no anthropological or linguistic legitimacy."[55] Karen Pennesi further contextualizes immigrants' name choices as "tension between the desire to belong, fueled by pressures to conform, and the desire to remain true to their own identities, mindful of linguistic and cultural origins as well as current social circumstances."[56] The onus of tolerance and acceptance here is on society rather than the migrants themselves, whose naming practices are presented as complex processes rather than measures of integration.

Oversimplifications such as the example above show the importance of a decolonial approach to naming practices. This becomes even clearer in Gerhards and Tuppat's description of their methodology, where they postulate that a name is considered "common" in Germany if the majority population (which is left undefined) holds this view.[57] They here include the name Kevin, explaining that it is not originally German but was introduced and popularized from the 1980s onward. However, they exclude names such as Mehmet and Mustafa, claiming that despite their growing popularity, they indicate immigrant background to the German majority population. This claim is clearly saturated with a bias toward northern and western European names and neglects to account for raciolinguistic ideologies, racism, and Islamophobia among the unspecified German majority population. It also emphasizes once more that a people-centered perspective is integral to decolonial thought because the Kevins, Mehmets, and Mustafas in question are not given any voice in this discussion. Finally, Gerhards and Tuppat's framing of naming decisions as "boundary-maintenance" or "boundary-crossing" shows how engrained the idea of names being bound to a certain language or community is even in academic discourse.

How can we disrupt this line of academic thought and its public reproductions? Linguistic anthropologists can engage in what Angela Reyes calls "postcolonial semiotics," processes meant to invoke "multiplicity, ambivalence, and uncertainty around the question of colonialism's endurance" in everyday signs.[58] Additionally, it is important to merge theory and practice, with both a focus on how contemporary events and politics shape our research as well as a commitment to impactful, accessible, and practical work.[59] As far as centering—and ideally improving—name bearers' lived experiences is concerned, there are many ways in which names scholars can make their work applicable and actionable for the broader public. Practical guidelines such as Pennesi (2017) or Pennesi and Guccini (2020) can help people treat unfamiliar names in their personal and professional environments in a more empathetic and evidence-informed manner. The references above include, for instance, recommendations for how to confirm and practice name pronunciations or set up inclusive online registration forms, calls to not make assumptions about someone's identity, appearance, and linguistic abilities based on their name, and advice on how to navigate and respect people's potential multiple and preferred names or nicknames. They also link to resources that people may use to further familiarize themselves with specific names, pronunciations, and names research.

Translanguaging Names Theory is not meant to be a catchall approach to decolonizing names, but it can help identify major drawbacks in the definitions and conceptualization of existing name theory. Names do not need to be universally translingual for an approach that recognizes potential translanguaging in all names to work. The crux of the argument is not that names do not have specific linguistic origins but rather that our categorization of those very origins is rooted in colonial ideology and therefore biased. A translanguaging framework also allows us to recognize naming practices that bridge and blur cultural and linguistic "boundaries." The meaning and reasoning behind such names are often lost on people whose attitude toward names adheres to monolingual bias. They might think a Sino-Mauritian family name "strange(ly long)" or ridicule Chinese students for choosing "weird English names," failing to recognize the colonial histories and ideologies that led to such translingual naming practices.

Conclusion

This chapter introduced an approach to onomastics, which I call Translanguaging Names Theory. This framework aims to decolonize language and naming practices by unraveling colonial ideologies of languages and language practices as separate systems and of names as inherently "belonging" to one language or community. The practice of translanguaging between and beyond rigid language categories challenges imagined boundaries between named languages, emphasizing instead a translingual repertoire from which speakers draw in manifold ways. Through a decolonial lens of translanguaging, we can then understand names as embedded in translingual repertoires rather than monolingual practices. Translanguaging Names Theory thus challenges preconceived notions of what makes a name "Dutch," 'Chinese," or "English," which exposes the raciolinguistic expectations we may have of people upon learning their name(s). Most importantly, for a decolonial perspective, we need to center people's lived experience of their names and name encounters so that we can understand translanguaging and naming as social practices that are impacted—but not determined—by colonial ideology.

Acknowledgments

I thank Drs. Lauren Beck and Grace A. Gomashie for the opportunity to contribute to this volume and their hard work on the editing process. I would also like to thank Dr. Karen Pennesi for many inspiring discussions about translanguaging and naming practices. My heartfelt gratitude to the participants of Case Studies A and B who shared their names and stories with me.

Notes

1 I interviewed Maia, a college student in the American Midwest, in January 2016 as part of Case Study B, one of the two datasets on which I am basing my argument.
2 Gabriela Veronelli, "A Coalitional Approach to Theorizing Decolonial Communication," *Hypatia* 31.2 (2016), pp. 404–420, 409. Also see Jonathan Rosa and Nelson Flores, "Unsettling Race and Language: Toward a Raciolinguistic Perspective," *Language in Society* 46.5 (2017), pp. 621–647, 623–624.
3 Valerie Alia, "Introduction. Towards a Theory of Political Onomastics," in *Names and Nunavut: Culture and Identity in Arctic Canada*, ed. Valerie Alia (New York: Berghahn Books, 2007), pp. 1–16, 8; Barbara Bodenhorn and Gabriele vom Bruck, "'Entangled in Histories': An Introduction to the Anthropology of Names and Naming," in *The Anthropology of Names and Naming*, eds. Gabriele vom Bruck and Barbara Bodenhorn (Cambridge: Cambridge University Press, 2006), pp. 1–30, 3.
4 Karen Pennesi, "'They Can Learn to Say My Name': Redistributing Responsibility for Integrating Immigrants to Canada," *Anthropologica* 58.1 (2016), pp. 46–59, 46.
5 Ofelia García and Wei Li, *Translanguaging: Language, Bilingualism and Education* (London: Palgrave Macmillan, 2014), p. 22.
6 Veronelli, "A Coalitional Approach to Theorizing Decolonial Communication," p. 407.
7 Federica Guccini, "Discourses of Tension in a Rainbow Nation: Transcultural Identity Formations among Hakka Mauritians." PhD dissertation, University of Western Ontario, 2022.
8 Federica Guccini, "A 'Sense of Recognition': Negotiating Naming Practices and Identities of Overseas Chinese Students in Transcultural Social Spaces," in *GISCA Occasional Papers* 12 (Göttingen: Institute for Social and Cultural Anthropology, 2017).
9 Ofelia García, Nelson Flores, Kate Seltzer, Wei Li, Ricardo Otheguy, and Jonathan Rosa, "Rejecting Abyssal Thinking in the Language and Education of Racialized Bilinguals: A Manifesto," *Critical Inquiry in Language Studies* 18.3 (2021), pp. 203–228, 206.
10 Robin Phelps-Ward and Wonjae Phillip Kim, "Centering Stories of Our Names and Systems of Oppression," in *The Power of Names in Identity and Oppression: Narratives for Equity in Higher Education and Student Affairs*, eds. Robin Phelps-Ward and Wonjae P. Kim (New York: Routledge, 2023), pp. 1–6, 1.
11 Robin Conley Riner and Rachel George, "Introduction," in *A New Companion to Linguistic Anthropology*, eds Alessandro Duranti, Rachel George, and Robin Conley Riner (Hoboken: Wiley-Blackwell, 2023), pp. 1–12, 8; emphasis in original.
12 Delmos J. Jones, "Epilogue," in *Decolonizing Anthropology: Moving Further Toward an Anthropology for Liberation*, ed. Faye V. Harrison (Arlington, VA: Association of Black Anthropologists, American Anthropological Association, 1997), pp. 192–200, 192.

13 Edmund T. Gordon, "Anthropology and Liberation," in *Decolonizing Anthropology: Moving Further toward an Anthropology for Liberation*, ed. Faye V. Harrison (Arlington, VA: Association of Black Anthropologists, American Anthropological Association, 1997), pp. 150–169, 155.
14 Jafari Sinclaire Allen and Ryan Cecil Jobson, "The Decolonizing Generation: (Race and) Theory in Anthropology Since the Eighties," *Current Anthropology* 57.2 (2016), pp.129–148, 136.
15 Catherine E. Walsh, "Decoloniality in/as Praxis," in *On Decoloniality: Concepts, Analytics, and Praxis*, eds. Walter D. Mignolo and Catherine E. Walsh (Durham, NC: Duke University Press, 2018), pp. 13–102, 81.
16 Rosa and Flores, "Unsettling Race and Language," pp. 621–647, 624. Also see Kathryn A. Woolard and Bambi B. Schieffelin, "Language Ideology," *Annual Review of Anthropology* 23 (1994), pp. 55–82, 67–68.
17 García et al., "Rejecting Abyssal Thinking," pp. 207–209, 213.
18 Veronelli, "A Coalitional Approach to Theorizing Decolonial Communication," p. 408. Emphasis in the original.
19 Aníbal Quijano, "Coloniality of Power and Eurocentrism in Latin America," *International Sociology* 15.2 (2000), pp. 215–232, 216.
20 H. Samy Alim, "Introducing Raciolinguistics: Racing Language and Languaging Race in Hyperracial Times," in *Raciolinguistics: How Language Shapes Our Ideas about Race*, eds. H. Samy Alim, John R. Rickford, and Arnetha F. Ball (New York: Oxford University Press, 2016), pp. 1–30; Nelson Flores and Jonathan Rosa, "Undoing Appropriateness: Raciolinguistic Ideologies and Language Diversity in Education," *Harvard Educational Review* 85.2 (2015), pp. 149–171; Rosa and Flores, "Unsettling Race and Language," pp. 90–107.
21 García et al., "Rejecting Abyssal Thinking in the Language and Education of Racialized Bilinguals," pp. 214–215.
22 Wei Li, "Moment Analysis and Translanguaging Space: Discursive Construction of Identities by Multilingual Chinese Youth in Britain," *Journal of Pragmatics* 43.5 (2011), pp.1222–1235, 1223.
23 A. Suresh Canagarajah, *Translingual Practice: Global Englishes and Cosmopolitan Relations* (London: Routledge, 2013), p. 6.
24 Mary Bucholtz, "On Being Called Out of One's Name: Indexical Bleaching as a Technique of Deracialization," in *Raciolinguistics: How Language Shapes Our Ideas About Race*, eds. H. Samy Alim, John R. Rickford, and Arnetha F. Ball (New York: Oxford University Press, 2016), pp. 273–289, 274.
25 Karen Pennesi, "Universal Design for Belonging: Living and Working with Diverse Personal Names," *Journal of Belonging, Identity, Language and Diversity* 1.1 (2017), pp. 25–44, 26.
26 Leslie Barratt, "Monolingual Bias," in *The TESOL Encyclopedia of English Language Teaching: Vol. 1*, ed. John I. Liontas (Oxford: Wiley-Blackwell, 2018), pp. 1–7, 1.
27 Canagarajah, *Translingual Practice*, p. 6.
28 García and Li, *Translanguaging*, p. 34.
29 Bodenhorn and vom Bruck, "'Entangled in Histories'," pp. 1–30, 2.
30 Pierre Bourdieu, *Language and Symbolic Power*, eds. John B. Thompson, trans. Gino Raymond and Matthew Adamson (Cambridge: Polity Press, 1991), p. 55.
31 Quijano, "Coloniality of Power and Eurocentrism in Latin America," p. 216.
32 Guccini, "Discourses of Tension in a Rainbow Nation."
33 Thomas Hylland Eriksen, *Common Denominators: Ethnicity, Nation-Building and Compromise in Mauritius* (Oxford: Berg, 1998), p. 4. Emphasis in original.
34 Clement Chan, *Hakkas Worldwide* (Moka, Mauritius: DCI Studios Ltd., 2010), p. 292.

35 There are, of course, exceptions to the rule even in China. The number of syllables may vary depending on regional context.
36 Ibrahim Alladin, *Philip Ah-Chuen: Entrepreneur, Philanthropist, Educator* (Coimbatore, India: Rukmani Offset Press, 2014), p. 38.
37 Research participants also reported using an alternate name in Chinese work environments. One of my interviewees insisted being called by her given name Qian during her studies abroad but had to choose an "English" name when she started an internship at a company in China. She appeared in their system as Joyce Q Li, her given name reduced to an initial and the order of her names following the "western" [first name] + [last name] style.
38 Man, Joyce, "Hong Kong Loves Weird English Names," *The Atlantic*, October 1, 2012; Morris, Ruth, "'Hello, My Name Is Yes,' and Other English Names in China," *Public Radio International*, April 22, 2014; Everington, Keoni, "Hilarious English Names Chinese People Choose." *The World of Chinese*, March 5, 2015; Brennan, Bo, "Chinese People's English Names—Is There a Method to the Madness?" *eChinacities*, 2015; Liu, Charles, "Laowai Entrepreneur Wants to Rid China of English 'Stripper Names,'" *The Nanfang*, 2015–04; Phillips, Tom, "US Entrepreneur Vows to Rid China of 'Farcical' Western Names: Lindsay Jernigan Hopes to Liberate Young Chinese from Bizarre Adopted Names Including Lady Gaga, Twinkle, Pussy, Elvis—and Washing Liquid," *The Telegraph*, March 31, 2015.
39 Bourdieu, *Language and Symbolic Power*, p. 57.
40 BestEnglishName.com. "BestEnglishName.Com." Retrieved September 7, 2016, http://www.bestenglishname.com.
41 BestEnglishName.com.
42 García et al., "Rejecting Abyssal Thinking in the Language and Education of Racialized Bilinguals," p. 209.
43 Paul McPherron, "'My Name Is Money': Name Choices and Global Identifications at a South-Chinese University," *Asia Pacific Journal of Education* 29.4 (2009), pp. 521–536, 527.
44 Laurie Duthie, "Western Names for Chinese Identities: The Acquisition and Use of Western Personal Names among Chinese Business Professionals in Foreign-Invested Corporations," *Asian Anthropology* 6.1 (2007), pp. 53–80, 66.
45 McPherron, "'My Name Is Money," p. 526.
46 Li, "Moment Analysis and Translanguaging Space," pp.1222–1235, 1227.
47 Li, "Moment Analysis and Translanguaging Space."
48 Bucholtz, "On Being Called Out of One's Name," pp. 273–289, 275.
49 Bucholtz, "On Being Called Out of One's Name," pp. 276–277; Karen Pennesi, "Differential Responses to Constraints on Naming Agency Among Indigenous Peoples and Immigrants in Canada," *Language & Communication* 64 (2019), pp. 91–103, 101.
50 Quijano, "Coloniality of Power and Eurocentrism in Latin America;" Veronelli, "A Coalitional Approach to Theorizing Decolonial Communication."
51 Angelique Chrisafis, "We Need to Talk about Kévin: French Namesakes Fight National Mockery," *The Guardian*, August 5, 2022.
52 Arthur K. Spears, "White Supremacy and Antiblackness: Theory and Lived Experience," *Journal of Linguistic Anthropology* 31.2 (2021), pp. 157–179, 159.
53 Bodenhorn and Bruck, "'Entangled in Histories'," p. 25.
54 Jürgen Gerhards and Julia Tuppat, "'Boundary-Maintenance' or 'Boundary-Crossing'? Name-Giving Practices Among Immigrants in Germany," *Names* 69.3 (2021), pp. 28–40, 30.
55 Jürgen Gerhards and Silke Hans, "Akkulturation und die Vergabe von Vornamen: Welche Namen wählen Migranten für ihre Kinder und warum?" in *Migration und*

Integration: Kölner Zeitschrift für Soziologie und Sozialpsychologie, Sonderheft 48, ed. Frank Kalter (Wiesbaden: Verlag für Sozialwissenschaften, 2008), pp. 465–487, 466–467; Federica Guccini, "A 'Sense of Recognition,'" p. 26.

56 Pennesi, "'They Can Learn to Say My Name'," pp. 46–59, 48.

57 Gerhards and Tuppat, "'Boundary-Maintenance' or 'Boundary-Crossing'?" pp. 31–32.

58 Angela Reyes, "Postcolonial Semiotics," in *A New Companion to Linguistic Anthropology*, eds. Alessandro Duranti, Rachel George, and Robin Conley Riner (Hoboken: Wiley-Blackwell, 2023), pp. 107–121, 108.

59 Riner and George, "Introduction," p. 4.

7 Aareck to Zsaneka

African American Nominals, Un/naming, and Aesthetic Justice

Jerrilyn McGregory

This chapter interrogates African American given names en route to developing their own self-affirmative and improvisational aesthetic. Symbology will be deployed as the premise from which to expound on the radically politicized turn toward un/naming. Because of the additional historiographic complexity of the original "unnaming," this study requires the backslash used in un/naming to distinguish today's neoteric nomenclatures from those of the western hegemonic nominal system. Situated as noisy names, this developmental shift within African American culture intimates the emergence of Victor Turner's ideation of a social drama underway. Driven by ritualized actions unique to rites of passage, in this case, racial injustice and socioeconomic oppression perpetuate the transgressive liberatory un/naming act that disregards any supposed marginalizing consequences. Arnold van Gennep first synthesized liminality as part of a social process with three stages: separation, liminality, and postliminal incorporation, each dependent on the larger society's abnegation—that is, awaiting ratification, nullification, and reintegration.[1] The post-Civil Rights period is seminal to the formation of a shifting identity among those then called Negroes, ultimately inducing an emergent Black identity formation or nigrescenece. Nigrescenece is a term articulated by psychologist William Cross to explicate identifiable stages among people struggling to embrace an alternative, previously disavowed marker of Black identity.[2]

An anomalous history significantly divested African Americans of their culturally rooted African anthroponyms, instead being obliged to be unnamed chiefly through the acquisition of western nomenclatures. So, nowadays, African American naming patterns defy this pejorative act of colonialism. Birth names from Aareck to Zsaneka developed expeditiously after centuries of enduring the imposition of other sociopolitical, cultural, and economic norms. An interesting canonical shift in African American naming patterns has accrued for more than a generation now. Most noticeably by 1980, African American birth names evolved, becoming more *sui generis*—maligned and adjudged to be pathological (if not humorous to the point of hilarity) by the power system. Demeaned as "ghetto," unique names exposed an improvisational expression of

DOI: 10.4324/9781003414773-8

militancy—preferable to perpetuating the reification of common western names. For my master's thesis, I culled the birth names of African American newborns in five-year intervals (1945–1980) from the Department of Vital Statistics in Gary, Indiana. Demographically, by 1980, African Americans constituted 71% of the residents there.[3] On the one hand, in that context and often bootstrapped to patronyms, male offspring ostensibly acquired comparatively more conservative nominals, with some of those uniquely multigenerational: Amon III, Evar Jr., and Sibley III. Female newborns, on the other hand, amassed a volley of link name formulations: Arlena, Carnida, Chaunel, Cheneel, Midinah, Starita, Tangela, and TiAngela.[4] These conclusions recall Dunkling's definition of link names as being "based on an existing name but add an extra element."[5]

Un/naming specifies nomenclatures that may not be aesthetically acceptable due to their disallowance of western onomastic signatures. This argument invests unnaming within African American vernacular English's standard formulation of semantic inversion, converting an expression into its opposite—to formulate a redemptive renaming.[6] As a matter of semantics, inversion relates to Ebonics or African American Vernacular English (AAVE), exemplars that rely on opposing expressions to encode meaning, as in "bad" to refer to something considered "good."[7] In essence, language is implemented to escape the world of order and constraint imposed by the dominant culture. Jordan Slump amplifies the same quandary:

> A belief in the concept of identity lies behind both an absolute rupture with tradition (in which one seeks to create and to impose one's own identity) and an absolute obedience to it (in which one identifies with it; one seeks to make oneself identical to it).[8]

In keeping with cultural dynamics, African Americans' significant rejection of dichotomous thinking allows for both considerations, too, with a mastery of an orthography of their own. As a matter of semantics, un/naming allows for subversive nomenclatures in keeping with what folklorist Zora Neale Hurston implicates as "the will to adorn, . . . the second most notable characteristic of Negro expression."[9]

In contrast, a stream of interdisciplinary onomastic research rely on quantitative methodologies consistent with fields such as Social, Clinical, and Developmental Psychology, Experimental Education, Labor Economics, among others. Inevitably, their conclusions invalidate African American names as pathological, leading to stereotypes using methods such as sociometric choice. This pervasive technique measures the personality development of children based on peer group belonging, in this case relating to names. After utilizing data measurements as an instrument to determine sociometric status, children's names are ranked by other youngsters and then categorized. Theoretically speaking, although not necessarily race-based, "positive sociometric choices [were] significantly related to the desirability of their first names."[10] Other quantitative studies privileging African American names and stereotypes engage with quotidian aspects of human existence such as employability, as well as rental and workplace discrimination.[11] The vast majority of quantitative analyses are casuistic and deploy prophetic summations that chastise oblivious name donors or, especially in the case of African Americans, level relentless opposition to onomastic creativity. Present-day un/naming practices steadfastly amplify and vilify nominal differences reliant on a discernible whiteness as an assumption underpinning this academic literature.

African Americans are not alone in comprising a marginalized group that push to have their names mainstreamed or recognized. Carmen Fernández-Juncal interrogates the evolution of onomastic usages since the transition from the Franco era, during which minority language groups were suppressed to Spain's democracy in the 1970s. She assessed the choice to grant children Basque names, such that Spanish ones become remarkably displaced by Basque names.[12] Likewise, French sociologist Baptiste Coulmont observed that the children of immigrants from North African continued to give their children Arab-Muslim names associated with North Africa. In the balance, however, some immigrants still reached for international or domestic names that promoted assimilation.[13] In an Ethiopian-American context, immigrant phenotype informed naming processes through symbolic interactionism, which, on one level, speaks to one's reference group orientation as well as the subjective meaning attached to one's epistemic status. The research concluded that immigrants chose two-syllabic names that are easy to pronounce and helped them avoid African American signifiers.[14]

The theory of internal colonialism denotes the domination, exploitation, and oppression that characterized the African American experience within U.S. borders. This political reality approximates the praxis of systemic domination experienced by non-white minorities in Australia and New Zealand, who encounter similar historical developments within a white settler colon. In a comparative study of African Americans and Māori families, R.A. Staples observes that "both groups coexist with European colonialists while suffering a devalued status based on skin color in their respective countries."[15] Regarding aboriginal people in Australia, and their dynamic perennial naming system, they garner nominal changes throughout life—dependent on contexts over a lifetime. In *Proof of Birth*, Rod Hagen advocates for balancing Indigenous processes for personal identification and those of the state and Australian settler society more broadly.[16] Many European countries still possess birth laws; France, for example, is a nation that for almost two centuries legislated and strictly adhered to a registry that delimited first name choices to prevent last names as first names and to protect children from potentially harmful ones until 1993.

The framing of African American un/naming as a decolonial practice clearly aligns domestic colonialism with cultural alienation informed by American history. Each political economy that historically has implicated African Americans—chattel slavery, tenant farming, rampant industrialism, and today's savage deindustrialization—constitutes a defining moment impacting African American name donors as the interlocutors of disparate naming customs. Traditional African names someone rooted their names to a local context that related somehow to shifting sites of renewal, memory, and retreat with a contextual-sensitive focus.[17]

Why, then, do African Americans denounce Anglicized proper names, inscribing stigmatized nomenclatures and rejecting Eurocentric nomenclatures equated with power and prestige? What aura might such quintessential signs of Blackness expel? What culturally gratifying prerequisite might such nominals satisfy? As it relates to aesthetics and decolonization, what initial role did Francophone creolization, particularly in Louisiana, play? Might the counterpoetics of these given names serve as abnegation and ensure social differentiation of African Americans should sociopolitical reintegration ever prevail? For reasons to be explored, prevailing indigence can best be articulated diachronically through the various lenses of the American political economy.

Nobody Knows My Name

As a part of his oeuvre, James Baldwin published four memoirs. Deemed a classic, *Nobody Knows My Name: A Letter from the South* entertains schemas depicting his first trip ever to U.S. southern states and his engagement with Martin Luther King during the Civil Rights period.[18] Baldwin's book title points to the complexity attached to his personal identity, his perceived evils of growing up as a second-class citizen, alongside his European experiences. A preacher in his youth, another of Baldwin's memoirs, *No Name in the Street,* quotes from Job 16–20. Taken together, these titles speak to a sense of displacement and spatial disembodiment that also invoke the imposition of other peoples' naming practices upon the African American presence in the New World.

Realizing that Africa is a vast continent, as are its naming practices, we envision the extent to which African American names stagnated due to adopting a predominantly western naming system. Over time, African names lost significance in an American location. Although African societies, too, function under certain strictures, a name's connotation remains of utmost importance; context is everything. Their systemic loss speaks generationally to the descendants of the men and women enchained aboard the White Lion, encapsulating the obdurate unnaming and the nominal erasure in the Americas, especially for the 20 or so Angolan captives who disembarked in 1619 on the Jamestown River in Virginia.[19] The White Lion sailed from a seaport in the Netherlands under a Dutch flag occupied by British corsairs who abducted captives from a Portuguese slave ship. Accordingly, the likelihood is that their "cargo" had already been Christianized by the Portuguese upon contact with Angolans over a century beforehand. To be indentured in Virginia in 1619 represented the bestowal of Anglicized cognomens along with the original unnaming of African peoples. Foremost, it signified for these Angolans the dispossession of their particularized Catholic identities among Virginian Protestants.[20] Folklorist Newbell Niles Puckett conjectured that, from the 11 known Anglicized appellatives, they moved to acquire Spanish sobriquets: "Thus, Anthony might be a metamorphized Anthonio, John may have emerged from Juan, William from Guillaume, Frances from Francisca, Edward from Eduardo, and Margaret from Margarita."[21] As part of the nascent transatlantic British slave trade, Puckett's revelation illustrates the fluidity of delegating involuntary nomenclatures onto Black bodies during the dawning of plantation colonialism in the United States.

By the end of the eighteenth century, some names shaped by the plantation system further dehumanized enslaved persons. From 1619 to 1799, the enslaved inherited the same hypocoristic names—Jack, Tom, and Harry—given to landowners' mules. Puckett goes on to observe that classical, pompous names such as Caesar, Cato, and Juno consistently formed a minute portion of names for those in bondage and among freedmen. In addition, descriptive, occupational, and titular names made appearances, but they, too never took hold. Moreover, it is questionable whether the so-called pompous names of African American patriots during the American Revolutionary War epitomized actual given names. In the balance, however, most names, regardless of gender, were of Biblical or African origins.[22] Other historical evidence indicates some retention of African names by both those manumitted as well as those in servitude.[23] For a compilation of the monikers of runaway enslaved people, advertisements for capture and return also facilitated onomastic research.[24] Nonetheless, the polemics popularized by Malcolm X's speeches comprise an important link to the condemnation that African Americans tolerated "slave names" without regard to the agency with which they most certainly humanized themselves, even under plantation colonialism.

Painstakingly and without a dependency on slaveowners' personal records, revisionist historian Herbert Gutman culled primary resources that demythologized prevailing rhetoric about slave names. Ostensibly published to disperse countervailing research that pathologized African American family units, Gutman substantiates that slaveowners possessed little compunction to confer names that cut across all segments of kinship, including parents, grandparents, great-grandparents, aunts, and uncles. While the majority possessed a common American given name, they "often carried it for reasons rooted in the developing Afro-American culture."[25] Also, influenced by West African customs, they selected posthumous names first given to a deceased sibling, and other data confirms the active role grandparents played in name decision-making. Then, too, the southern aristocracy routinely bestowed upon the daughter her mother's name, a matrilineal practice not adhered to by those enslaved. Consistent with bondage and the likelihood of marriages being disrupted at auction, the firstborn son aptly received patronyms that bolstered a dyadic relationship because fathers were more likely to be sold or to suffer a premature death. Gutman attributes this practice to the probability that plantation owners seldom recorded the names of slave fathers, only recognizing uterine descent by notating the mother's appellation. In this sense, typically slave parents bestowed names upon their children, although there is evidence that slaveholders also named the children of the enslaved.[26]

From Nominal Freedom to Proletariat

With the end of the antebellum period after the Civil War, Reconstruction unleashed a feudalistic emancipation crisis that eventually initiated an out-migration of African Americans, known as the Great Migration. First, sharecropping constituted the only path forward for restricted freed persons; the political economy was tantamount to enslavement that hinged on exasperating this landless peasant class. The South's shift in political economies resulted in a tenant farming system within which laborers seldom owned the land they cultivated. During Reconstruction, the Freedmen's Bureau functioned as the first federal welfare agency, performing the Herculean task of creating a bridge from slavery to freedom. The Bureau compiled onomastic evidence indicating that those newly emancipated resisted the retention of names associated with their slave masters. Those formerly enslaved had to undertake a task unknown to free-born Americans when the Bureau required them to adopt last names. For example, variants of the surname Freeman emerged, communicating a quest for new identities for themselves and future generations. Although enslaved persons often assumed last names for use among themselves, few masters condoned bestowing the dignity of their surnames upon chattel. In 1935, during the Great Depression, President Roosevelt established the Works Progress Administration, which collected narratives from enslaved antebellum survivors whose agentive discourse included personal experiential evidence regarding the adoption of surnames. In their personalized testimonials, they exposed a variety of rationales for toppling the surnames of the aristocracy as well as alternative rationales, if otherwise.

These accounts often emphasized farcical determining factors for the serial procurement of surnames, dependent on ever-shifting esoteric home contexts. For example, Prince Johnson's in-depth qualitative narrative reveals how the marriage of a slaveholder's offspring dismantled the lives of hundreds of enslaved persons and customarily their claimed surnames: "Den us names was changed from Love to Johnson, and prior to this dey names was Dennis when dey come, but, after de custom o dem days, dey took de name o Love from dey new owner."[27] To procure citizenship rights, government bureaucracies hastily demanded identification: "We had to register as someone, well, I got to thinking about all us slaves that was going to take the name Fitzpatrick. I made up my mind I'd find me a different one. One of my grandfathers in Africa was called Jeaceo, and so I decided to be Jackson."[28] Rather than sycophantic affection for their owners, it was only out of necessity that those newly emancipated deemed it the logical and easiest way to proceed and gain citizenship. James Lucas, bearing the distinction of having been enslaved by the Confederacy president, Jefferson

Davis, and with rare awareness of his own birth date, October 11, 1833, imparts an incredible memory of his family's genealogy. He references his grandmother who "come over to dis country in a big ship and folks couldn't un'erstan a word she say." Although given the name Betty, she insisted that it was not her "right name."[29]

As sharecroppers, their agrarian reality chiefly constituted the main consolation prize after generations of bondage. The institutionalization of Jim Crow (Black codes) evolved structurally to ensure the reign of white supremacy, with segregation always accompanying internal colonialism. Politicians enjoined state laws to enforce supposedly "separate but equal" accommodations for African Americans and whites to the absurd extent of these laws forcibly separating them along racial lines when it came to public transportation, sports, hospitals, orphanages, prisons, asylums, funeral homes, morgues, and cemeteries. Historian Arnold Taylor asserts as by the 1930s, Black farmers had gained little ground in the South compared to the period leading up to Emancipation.[30] Fin de siècle, indicative of the far-reach of structural racism endured by the first wave of the Great Migration northward, mainstream, and academic publications, situated "colored names" uncritically into stereotypical, esoteric curios. *The New York Times Magazine*, for instance, published articles such as Frances Gaither's "Fanciful Are Negro Names: Meet Artificial Flowers, Quo Vadis, Parolee and the Twins Avery and Ivory," adjudging them to be not only humorous but also as plain obstacles to social advancement.[31] In 1929, Naomi Chappell lambasted Negro names as an old and peculiar fancy representative of an ear for rhythmic or euphonious names. She listed 15 examples retrieved from the author's maid: Hilda Vernetta, Antoinette Alva, Wildetta Allien, Deodolphus Douglas, and Dexter Poindexter. Beyond being judged merely ridiculous and pathological, Chappell's article indulges commonplace racial stereotypes intimating that those with puritanical, resonant names like Comfort, Hallowed, Ocean, and Memory were in search of absolution for their criminal propensities. Even the bestowal of common names such as Lily and Pearl incited degradation as being "sympathetic magic, to lighten the dusky hue of some dark-skinned maiden."[32] Gaither's article unceremoniously concludes by enumerating examples of those allegedly possessing four or more "baptismal names," such as a chauffeur named Theophilus Otis Israbestis Tott and that of a colored minister's son, Matthew Mark Luke John The Acts of the Apostles Son of Zebedee Garden of Gethsemane Hill. In pursuit of financial opportunities, additionally, the offspring of migrants often retained names deemed countrified up North where African American elites shared many cultural values with the members of larger society.[33]

Know the Name

Born between 1946 and 1964, a baby boom occurred when older couples commenced bearing children in conjunction with younger ones, establishing an expansive increase in the birthrate. As Landon Jones articulates, "the fertility rate of African Americans closely approximated that of most other Americans," despite grievous differences in social and economic circumstances.[34] Alluding to the post-Second World War second migratory wave (known as "the silent arrival"), another avalanche of African Americans migrated from the South northward and beyond with great expectations. Their names often reflected a relatively antiquated quality such as Emmett and Evelene, Conroy and Erma, or Ulice and Estella.[35] African American names divulge no small share of details about the anthroponyms spawned during this period of northern industrialization with modern capitalism on the rise. Historically, and although they bestowed a wider variety of names to offspring, these were not necessarily unusual or odd and began to succumb to names consistent with the white populous. In the midst of the postwar baby boom, African Americans' sense of prosperity profoundly exhibited their projection of economic stability within a highly industrialized America.

Hindsight reveals that the names parents privileged synchronized with their economic advancement. Considered old-fashioned nowadays, their choices amplified the political economy of this particular defining moment. For instance, in 1945, popular girls' names included Barbara, Gwendolyn, Gloria, Patricia, Sandra, Dorothy, Joyce, Shirley, Carol, Donna, and Yvonne, and for boys (other than Willie), Ronald, Robert, James, John, Charles, William, George, Raymond, Percy, and Donald. As Meier and Rudwick conclude, "At midcentury the Great Migration to the industrial centers since the First World War was about to usher in the most momentous changes in black-white relationships since the Civil War."[36] This supports a central contention that the years leading up to the Civil Rights movement generated a strong sense of financial stability among northern African American parents who often benefited from conforming to the dual-earner model.

On the one hand, adopting names similar to their white counterparts accentuated the aesthetic and paradigmatic shift toward un/naming in the latter part of the twentieth century. African Americans embraced an urban lifestyle that challenged the status quo by sustaining their own cultural and aesthetic survival to play out the truths of their contemporary life experiences and not repeat what had been done by the generations before them.[37] The proper names of girls typically demonstrate wider diversification; however, it is the broadening of African American male names that becomes the greatest harbinger of change. By 1965, the proper names of African American children began to experience a transformation. Mothers born in Indiana contributed over half of newborn girls' unique link names, while male names incorporated suffixes such as -ell (Darnell, Rondell, Jerriel) and prefixes Da-, De-, and La- (Dashawn, DeAngelo, and LaDurian).[38]

Society began to take notice. In 1968, *Newsweek* published the article titled "Black Names," being one of the first exposés to recognize a national onomastic shift as signs of political disaffection.[39] This leads naming patterns along a trajectory toward its present state of being, un/naming.

No-Name

En route to developing their own self-affirmative, improvisational aesthetic, having "no-name" points to non-Black society's view of Black names, while African Americans delight in *sui generis* names. When I queried Jazzime Gainour, my dental hygienist, about her name, her first words were, "We like to have our fancy names!" Epitomizing the long march in search of redressive measures and eluding dichotomous thinking, un/naming is indicative of an ongoing political rupture in American society. In the aftermath of the decolonizing rhetoric of Black Cultural Nationalism, children born thereafter continued to be allocated neoteric names that fashioned greater aesthetic distance from western hegemonic names. African American name givers reclaimed, if not African names, a heightened array of euphonymic ones that are sensuous when spoken. Although mainstream politics paid lip service to redressive measures such as Affirmative Action and recently provoking the Black Lives Matter movement, true social justice has yet to materialize. In keeping with Turner's social drama, then, a preponderance of evidence betokens the perpetuation of marginality as a politicized state of crisis without any expectation of reintegration or abrogation.

It is noteworthy also to mention the late bell hooks' use of liberating subjectivity. As a public intellectual, hooks devoutly dedicated her career to exposing the systems that oppress and dominate the marginalized. Regarding the Black Power Movement, she addresses its identity politics thus: "Gone was the notion that bonding with blackness was a one 'approved' black identity by how one responded to whiteness."[40] Ultimately, hooks' stance speaks to Black nationalism's tendency to bifurcate an essentialist and patriarchal theorizing of Black subjectivity. Instead, hooks ascribes to the formation of a liberating subjectivity: "Fluidity means that our black identities are constantly changing as we respond to circumstances in our families and communities of origin, and as we interact with a larger world."[41] Naming patterns are tied to his fluidity and this shift in naming patterns represents one of the only remaining vestiges of the 1960s in its aftermath. TRIOS is an acronym referencing time, rhythm, improvisation, orality, and spirituality as core psychological dimensions of African American culture. Acronyms such as TRIOS support a dialectical viewpoint "to create an integrated collective and individual-level worldview."[42]

In essence, present-day African American nomenclatures have grown beyond being "Black" names. The popularizing of negative associations with so-called Black Names implicates Bill Cosby's notorious 2004 "Pound Cake" speech along with the 2005 publication of the bestseller *Freakonomics: A Rogue Economist Explores the Hidden Side of Everything* by Stephen Dubner and Steven Levitt. Both phenomena unleashed an avalanche of pathological suppositions regarding African American nomenclature via mass media. Even prior to Cosby's criminal woes, this speech propagated a transgenerational firestorm, ostensibly revealing the depth of the social class divide. His oration on the occasion of the

NAACP's Legal Defence Awards Program to commemorate the fiftieth anniversary of the *Brown vs. Board of Education* decision garnered approving accolades from its designated audience; however, the following woeful sentiments were rebuked elsewhere:

> What part of Africa did this come from? (laughter). We are not Africans. Those people, not Africans, they don't know a damned thing about Africa. With names like Shaniqua, Shaligua, Mohammed, and all that crap and all of them are in jail. (When we give these kinds of names to our children, we give them the strength and inspiration in the meaning of those names. What's the point of giving them strong names if there is no parenting and values backing it up?

As one of his most robust critics, sociologist, and public intellectual Michael Eric Dyson concludes, Cosby's rhetoric coincides with the African American cult of respectability.[43] Indeed, African Americans are not monolithic and some, in the case of Crosby, engage in forms of transgenerational class warfare with a propensity to demonize the working class by adopting many of the ideologies of wider, non-Black society.

Due to clever marketing, the commercial success of *Freakonomics: A Rogue Economist Explores the Hidden Side of Everything,* written by highly established academics Steven Levitt and Stephen Dubner, commodifies African American naming trends. The chapter "Perfect Parenting, Part ii: Would a Roshanda by any Other Name Smell as Sweet" charts the whitest and blackest names to determine whether "the name you give your child affect his life? Or is it your life reflected in his name? In either case, what kind of signal does a child's name send to the world—and most importantly, does it really matter?" While more nuanced than the audit studies upon which its enumerative lists depend, *Freakonomics* deploys a rhetorical strategy that possibly contributed to the reframing of the stuff African American urban legends depend on. The "Roshanda" chapter concludes:

> But the kind of parents who name their son Jake don't tend to live in the same neighborhoods or share economic circumstances with the kind of parents who name their son, DeShawn. And that's why, on average, a boy named Jake will tend to earn more money and get more education than a boy named DeShawn.[44]

Someone Knows My Name

Critical race aesthetics is a burgeoning area of philosophy applicable to implicit racial bias toward so-called African American-sounding names that fall outside the Anglophile onomastic tradition. They evoke a transmogrifying un/naming aesthetic—the strategic rejection of conformity to hegemonic names vis-à-vis onomastic nationalism and racial supremacism. African American *sui generis* personal name choices reject the demands imposed by social stratification's vertical hierarchy. Like the rise of Afro hairstyles as marking a Black aesthetic turn, names form another projection of this aesthetic. Addison Gayle, Jr. articulates this phenomenon in terms of

> the acceptance of the phrase "Black is Beautiful" is the first step in the destruction of the old table of the laws and the construction of new ones, for the phrase flies in the face of the whole ethos of the white aesthetic.[45]

Whereas civil disobedience over present-day political injustices is visually accessible on daily news programming, nominal disobedience is denoted by the perpetuation of nonhegemonic African American birth names. Nominal disobedience signifies that for more than a half-century now, African American so-called fancy names flourish, becoming even more radicalized and improvisational as a political and cultural mainstay. I have noticed the given names of recent African American game show contestants on "Family Feud" as well as the 2024 NFL draft, both of which pinpoint the common representation of this alternative nominal system:

Projecting aesthetic injustice to be a nemesis to democracy, philosopher Gustavo Dalaqua argues that "if the oppressed refuse to believe in the social hierarchy invented by the oppressor to demean them, oppression cannot sustain itself."[46]

Coinciding with hip-hop culture's ascendancy, African American naming patterns echo and reverberate in sync with and integral to the production of new stylistic sensibilities. As Murray Forman argues:

> The hood is constituted as a privileged realm of interaction and emerges through its connections to the locality as a new 'scale of analysis' for the definition of experience and identity in a world of diminishing opportunities and nonsustaining political and economic structures among minority youth.[47]

The combination of locality and identity within hip-hop culture pivots from the periphery to center on the awakening of ghettocentricity: "It attracts listeners

Table 7.1 Comparison of African American names, Family Feud and NFL draft, 2024.

Family Feud Sisters	*NFL Draftees*
Karnetria	Kamari
Kaytisha	Ja'Tavion
Karleshia	Ga'Quincy
Karmishala	Qwan'tez
Karprice	T'Vondre

for whom the 'ghetto' is a place of adventure, unbridled violence, erotic fantasy, and/or an imaginary alternative to suburban boredom."[48] Therefore, coinciding with hip-hop culture's ascendancy, African American naming patterns echo and reverberate with an aesthetic movements in recent years. Aesthetic injustice appertains to when the fashion, taste, or aesthetic choices of a subaltern group are marginalized because they do not coincide with normalized standards. Such differences project dire economic consequences for African Americans who refuse to assimilate, navigating away from hegemonic nomenclatures and preferring personal names derived from their own aesthetic imagination. Reconstructed appellations function as a canonical shift toward un/naming, decolonizing personal names via the unapologetic distillation and emancipation from the power of the white gaze.

Conclusion

This chapter has engaged with how the dialectical law of hegemonic similarity exhibits rhetorical racism, illustrating cause and effect through the individual act of naming one's offspring and producing a nominal act of self-decolonization. Unlike global decolonization elsewhere, African Americans' inability to revert to traditional Indigenous spaces and societies of Africa symbolizes how their neoteric naming patterns bespeak an objection to utopian fantasies. Also, African Americans are not one homogeneous, monolithic community; yet, foremost, their given names serve as a barometer of ongoing deindustrialization, disenfranchisement, and their cries to defund the police. The abandonment of many popular names of girls from the 1950s, like Brenda, Cynthia, and Mary, to more novel ones, such as Qui'ana, Nakia, and Tamika, are indicative of aesthetic disobedience. Male names may be more patrilineally bound and, therefore, seem more common; yet they often bear witness to a preponderance of rarefied apostrophes as accent marks such as Ta'lon, D'Maurian, and Tre'. Such names remain a primary cultural vestige from the post-Civil Rights era, whereas Christian societies often curtailed the growth of acceptable nominatives by requiring children to be named for saints or strictly from the Bible. Over time, those sanctioned names have lost significance. W.F.H. Nicolaisen indicated that despite the inferred "connotative potential" of names, they possess major denotative functions, too. Martha S. Jones helps to define the dynamics afoot to adjudicate African American current naming patterns:

> Naming is one essence of freedom. Some enslaved people resisted subjugation when naming their children; some changed their names, hoping to elude greedy owners and brutal slave catchers. With emancipation, many more threw off the names given to them by slaveholders, acquiring for the first time last names such as Freeman that passed on how it felt to savor the first moments of liberty. Even today, some of us carry the names of the families

who called our forebears property. Also among us are those who are called by "X" and by unique names, signs of how the quest for freedom persists.[49]

Consequently, un/naming garners a negative connotation not only because it evades hegemonic onomastic structures to the child's detriment but also because the decolonizing act functions as a ritualistic and symbolic gesture to attain socio-political and aesthetic justice. And what if, eventually, this liberatory act follows the usual pattern of cultural appropriation pattern and is consumed by mainstream Americans?

Notes

1 I center the ritual process established by Arnold van Gennep and Victor Turner as pertinent to this study due to the structural racism that situates African Americans (onomastically speaking) "betwixt-and-between," by dramatizing the decolonizing of their children's names. See van Gennep's *The Rites of Passage* (Chicago: University of Chicago Press, 1960) along with Turner's *The Ritual Process and Dramas, Fields, and Metaphors* (Ithaca, NY: Cornell University Press, 1974) for greater elaboration.

2 Nigresence is the terminology introduced by psychologist William Cross, who devised five stages in the development of "new" Black identity formation: (1) Pre-encounter; (2) Encounter; (3) Immersion-Emersion; (4) Internalization; and (5) Internalization-Commitment. Internalization is the stage in which some break rank as is now the case regarding the adoption of African American. See William Cross, "Models of Psycological Nigrescene: A Literary Review," in *Black Psychology*, ed. Reginal Jones (London: Harper & Row, 1980), pp. 84–90.

3 U.S. Department of Commerce, *Bureau to the Census County and City Data Book* (Washington, DC U.S. Government Printing Office, 1983), p. 710.

4 Jerrilyn McGregory, "Aareck to Zsaneka: African American Names in an Urban Community, 1945–1980," Master's thesis, Cornell University, 1985.

5 Leslie Alan Dunkling, *First Names First* (Detroit, MI: Gale Research, 1977), p. 110.

6 See John R. Rickford, *African American Vernacular English: Features, Evolution, Educational Implications* (London: Wiley, 1999). Joey Lee Dillard is one of the first linguists to comprehensively convey that, rather than being deviant, Black English is a language in its own right that informs African American naming practices. See Joey Lee Dillard, "On the Grammar of Afro-American Naming Practices," *Names* 16.3 (1968), pp. 230–237, and by the same author, *Black English* (New York: Vintage Books, 1972).

7 Geneva Smitherman, *Talkin That Talk: Language, Culture, and Education in African America* (New York: Routledge, 2000), p. 21.

8 Jordan Stump, *Naming & Unnaming: On Raymond Queneau* (Lincoln: University of Nebraska Press, 1998), p. 165.

9 Zora Neale Hurston, "Characteristics of Negro Expression," in *Signifyin(g), Sanctifyin', and Slam Dunking*, ed. Gena Dagel Caponi (Boston: University of Massachusetts Press, 1999), pp. 293–308, 294.

10 Louisa Seraydarian and Thomas V. Busse, "First-Name Stereotypes and Essay Grading," *Journal of Psychology* 108.2 (1981), pp. 253–257.

11 M. Bertrand and S. Mullainathan, "Are Emily and Greg More Employable than Lakisha and Jamal? A Field Experiment on Labor Market Discrimination," *American*

Economic Review 94.4 (2004), pp. 991–1013; Kendra Hamilton, “What’s in a Name? Study Shows That Workplace Discrimination Begins Long before the Job Seeker Shows Up for an Interview,” *Black Issues in Higher Education* 20.9 (2003), p. 28; Adrian Carpuser et al., “Rental Discrimination and Ethnicity in Names,” *Journal of Applied Social Psychology* 36.4 (2006), pp. 934–952.

12 Carmen Fernández-Juncal, “Basque Evolution of Anthroponyms in an Area of Linguistic Transition: A Socio-Onomastic Study,” *Names* 66.2 (2018), pp. 85–95.

13 Baptiste Coulmont, *Sociology of First Names* (Paris: La Découverte, 2015). Also see Baptiste Coulmont and Patrick Simon, “How Do Immigrants Name Their Children in France?” *Population et sociétés* 565 (2019), pp. 1–4.

14 Hewan Girma, “Black Names, Immigrant Names: Navigating Race and Ethnicity Through Personal Names,” *Journal of Black Studies* 51.1 (2020), pp. 16–36.

15 R. A. Staples, “A Comparative Analysis of Afro-American/Maori Families: The Subtext of Minority Status,” *The ABNF Journal* 4.1 (1993), pp. 5–9. Also see Margaret Wetherell and Jonathan Potter, *Mapping the Language of Racism: Discourse and the Legitimation of Exploitation* (New York: Columbia University Press, 1993).

16 Rod Hagan, “Traditional Australian Aboriginal Naming Processes,” in *Proof of Birth*, eds. Melissa Castan and Paula Gerber (Albert Park, Victoria: Future Leaders, 2015), pp. 87–101, 99.

17 Examining many cultures across Africa, Sambulo Ndlovu’s *Naming and Othering in Africa: Imagining Supremacy and Inferiority Through Language* bolsters a comprehension of eclectic nominal fealty applicable to both those deemed contextually positive or negative.

18 Eddie S. Glaude, *Begin again: James Baldwin’s America and Its Urgent Lessons for Our Own* (New York: Crown, 2020).

19 Publication of *The 1619 Project* honored the 400th anniversary of the arrival of twenty or so Angolan captives in colonial Virginia. https://www.nps.gov/coro/learn/historyculture/esteban-de-dorantes.htm.

20 Dauril Alden, *The Making of an Enterprise: the Society of Jesus in Portugal, its Empire, and beyond: 1540–1750* (Stanford, CA: Stanford University Press, 1996).

21 Newbell Niles Puckett, *Black Names in America: Origins and Usage* (Boston: G.K. Hall, 1975), p. 7.

22 Puckett, *Black Names*, pp. 10–11, 45. Among the many Negro soldiers who fought at Bunker Hill were Prince Hall, Pomp Fisk, Cuff Hayes, Caesar Dickerson, Caesar Weatherbee, and Salem Poor. Lerone Bennett, *Before the Mayflower* (Baltimore, MD: Penguin, 1968), p. 43.

23 Those manumitted also preferred Old Testament names and African names gained popularity as a type of status symbol. See Puckett, *Black Names*, p. 46; Hennig Cohen, “Slave Names in Colonial South Carolina,” *American Speech* 27.2 (1952), pp. 102–107; J. C. Inscoe, “Carolina Slave Names: An Index to Acculturation,” *The Journal of Southern History*, 49.4(1983), pp. 527–554.

24 Relying on such ads as material resources, for example, Laversuch aggregated a typology comprised of eight categories: (1) plants and animals; (2) locations; (3) the Bible; (4) classical; (5) aristocracy; (6) African; (7) occupations; and (8) surnames. I. M. Laversuch, “Runaway Slave Names Recaptured: An Investigation of the Personal First Names of Fugitive Slaves Advertised in the Virginia Gazette between 1736 and 1776,” *Names* 54.4 (2006), pp. 331–362, 340.

25 Herbert Gutman, *The Black Family in Slavery and Freedom, 1750–1925* (New York: Vintage, 1976), p. 53.

26 Eugene Genovese, *Roll Jordan Roll: The World the Slaves Made* (New York: Vintage, 1976), p. 450.

27 Federal Writers Project, *Mississippi Slave Narratives: A Folk History of Slavery in Mississippi from Interviews with Former Slaves* (Bedford, MA: Applewood Books, 2006), p. 76.
28 "We was free. Just like that, we was free." National Humanities Center, 2009: nationalhumanitiescenter.org Emancipation through Union occupation and victory, selections from the WPA narratives, 1930s (nationalhumanitiescenter.org)
29 Federal Writers Project, *Mississippi Slave Narratives*, p. 91.
30 Arnold Taylor, *Travail and Triumph: Black Life and Culture in the South Since the Civil War* (Westport, CT: Greenwood Press, 1976).
31 Frances Gaither, "Fanciful Are Negro Names; Meet Artificial Flowers, Quo Vadis, Parolee and the Twins Avery and Ivory," *New York Times Magazine*, February 10, 1929.
32 Puckett, *Black Names*, pp. 274–275.
33 Sterling Stuckey, *Slave Culture: Nationalist Theory & the Foundations of Black America* (New York: Oxford University Press, 1987), p. 232.
34 Some believe this generation resulted from returning veterans after the Second World War. See Landon Jones, *Great Expectations: America and the Baby Boom Generation* (New York: Coward, McCann & Geohegan, 1980), p. 33. Also see S. Preston et al., *The Demography of African Americans 1930–1990* (Berlin: Springer Netherlands, 2013), p. 143.
35 McGregory, "Aareck to Zsaneka," 1985.
36 August Meier and Elliott M. Rudwick, *From Plantation to Ghetto: An Interpretive History of American Negroes* (New York: Hill and Wang, 1966), pp. 58, 270.
37 Salaam, Kalamu, "It Didn't Jes Grew: The Social and Aesthetic Significance of African American Music," *African American Review* 29.2 (1995), pp. 351–375, 370; Robin Kelley, *Race Rebels: Culture, Politics, and the Black Working Class* (New York: Free Press, 1996), p. 440.
38 McGregory, "Aareck to Zsaneka," p. 123.
39 "Black Names," *Newsweek* 62 (July 29, 1968), p. 80.
40 bell hooks, *Killing Rage: Ending Racism* (New York: H. Holt and Co., 1996), p. 241.
41 hooks, *Killing Rage*, p. 250.
42 James Jones, "TRIOS: A Model for Coping with the Universal Context of Racism," in *Racial Identity in Context: The Legacy of Kenneth B. Clark*, ed. Gina Philogène (Washington, DC: American Psychological Association, 2004), pp. 161–190, 161.
43 Michael Eric Dyson, *Is Bill Cosby Right? Or Has the Black Middle Class Lost Its Mind?* (New York: Basic Civitas, 2005), xv.
44 Steven Levitt and Stephen Dubner, *Freakonomics: A Rogue Economist Explores the Hidden Side of Everything* (New York: HarperCollins, 2005), p. 173.
45 Addison Gayle, Jr., "Cultural Strangulation: Black Literature and the White Aesthetic," in *The Black Aesthetic*, ed. Addison Gayle, Jr. (Garden City, NJ: Doubleday, Garden City, 1971), p. 40.
46 Gustavo Dalaqua, "Aesthetic Injustice," *Journal of Aesthetics & Culture* 12.1 (2020), article 1712183, p. 3.
47 Murray Forman, *The 'Hood Comes First: Race, Space, and Place in Rap and Hip-Hop* (Middletown, CT: Wesleyan University Press, 2002), p. 63.
48 Kelley, *Race Rebels*, p. 191.
49 Marsha S, Jones, "Ida, Maya, Rosa, Harriet: The Power in Our Names," *The New York Times*, June 18, 2020, https://www.nytimes.com/2020/06/18/style/self-care/sojourner-truth-harriet-tubman-slavery-names.html?auth=login-google1tap&login=google1tap

8 Same Old, Same Old

How Postcolonialism Didn't Change Things in Singapore

Peter K. W. Tan

John Stuart Mill argued that names (proper nouns) are to be distinguished from common nouns in that names are non-connotative, and the function of names is only to refer.[1] There is no meaning behind names. This position appears to have been re-affirmed by Kripke, who argued that names function as rigid designators in all possible worlds and circumstances. Coates, in teasing out the notion of senselessness associated with names, goes on to say,

> But even if names are senseless, it does not follow that they are meaningless; they may be used to refer, and if one wishes to express it so, the meaning is the referent. . .; they may set up cultural expectations. . . . And carry a great deal of social information; they may have transparent etymologies.[2]

These cultural and social elements need to be picked up in our discussion of postcolonial names.

The aim of postcolonialism is "to question, dismantle and subvert the binary hegemonic constructions that have supported and justified colonialism."[3] We can imagine names bestowed in the colonial period to places, ranging from towns and cities to street names and buildings. These names will establish particular priorities. Some of these priorities can be seen in linguistic terms: one way of inscribing colonial status to the landscape is to populate it with names in the colonial language. In writing, this would mean representing the name with the script associated with the colonial language (and in the case of English, this would be the Latin script), as well as the convention of using that script in that language. If the name contains a generic element (that is, words that refer to the nature of the entity being named, such as a street or a bridge), that would be in the colonial language. Finally, if the name itself is descriptive in a general way (for instance, Church Street is named after a church in the area), then the descriptive word would also be in the colonial language. Some of this can also be seen in the choice of commemorative names if that particular style of naming is employed. Commemorative names are given to honor especially particular

DOI: 10.4324/9781003414773-9

individuals and sometimes also places and events. The names thus inscribed into the place names indicate the priorities of the government of the time. Beyond the public naming practices, we might also consider naming in the private sphere, including personal names, business and commercial names, and names in literature and the arts.[4]

The enterprise of changing or reversing these colonial influences in names in postcolonial societies is often taken up by the new governments with new languages and new heroes to promote. As such, this could be seen as something under the aegis of planning and policy (as has been done in another chapter, for instance), and these might be challenged by forces such as globalization and commercialization (as I have discussed elsewhere).[5] Clear examples of this phenomenon would be place and street name changes such as has happened in Kuala Lumpur, for instance, Swettenham Road is now Jalan Sultan Salahuddin. The language has changed, so the generic element is now in Malay (*Jalan*) rather than in English (*Road*). And the person being commemorated has changed too.[6] An example of a spelling change would be that of *Melaka* (the city and the state), where previously *Malacca* was the norm in English. The former spelling conforms to how the Latin alphabet is used in Malay (though we might argue that the fact that Malay is written in the Latin alphabet itself shows colonial influence).

This chapter focuses on names in Singapore, and I will cast my net wide to include a range of different names found there, all collected from earlier research. The notion of top-down as opposed to bottom-up names discussed by Landry and Bourhis is useful to distinguish between names given by government agencies and those that are not.[7] Top-down names can work in concert with bottom-up names, and in some places they do not, as is the case for Malaysia.[8] The Singaporean names selected for examination encompass this range of name types. I will also attempt to account for the naming patterns found. The title of this chapter already hints at the treatment given to names in postcolonial Singapore. The most obvious way of signaling the postcolonial status is to adopt a *volte-face* in a range of matters, including naming. This gets a lot of attention, perhaps even encompassing an element of bravura. A less obvious strategy might involve retaining some of the structures erected by the colonial government because these structures can serve the needs of the postcolonial nation and only require tweaking rather than wholesale replacement. To the untrained eye, it might *appear* that nothing has changed. The old can, however, be re-purposed. The strategy might be a more self-effacing one. In the analysis below, I hope to show how the more low-key route might indeed be what Singapore has gone for. Key to the consideration are two views: that it is important that names reflect and not erase the history of Singapore and that the English language does not just belong to the British. These will be developed later on in the chapter.

Background to Singapore

Today, Singapore is a city-state with a population of 5.45 million, as given by the Department of Statistics.[9] Thomas Stamford Raffles founded a base for the British here in 1819. At that time, Singapore was inhabited by "sea gypsies" (*orang laut*) in fishing villages. Its founding drew many immigrants from China and elsewhere so that it became a base for Chinese and British economic interests in the region. With Penang and Malacca, Singapore formed the Straits Settlement in 1826 and was transferred to the Colonial Office in 1867. Today some three-quarters of the population are of Chinese origin. The fall of Malaya to the Japanese in the Second World War led to the erosion of British power and eventually to independence being given to Malaya in 1957. Singapore joined Malaya and other states to form the Federation of Malaya in 1963 but was thrust out of the Federation in 1965.[10] In Malaysia, the only official language is Malay, whereas in Singapore, there are four official languages: Malay, English, Mandarin Chinese, and Tamil. English is designated the "working language," the main language in schools, and has also become the main home language. As we are concerned about names, it is pertinent to say that "Singapore" is the Anglicized version of the Sanskrit name Singapura (*singa*, lion; *pura*, city).[11] The Romanized Sanskrit form has been retained in the Malay version of the name, but the English form was established in the nineteenth century, and this is the form that continues to be used. During the Japanese Occupation (1942–1945), the name was changed to Syonan-to, meaning "light of the south" or "the prosperous south," but reverted to Singapore when the Japanese left.[12]

In a novel that considers an alternative history of Singapore, Kevin Martens Wong's *Altered Straits*, we are given a description of the Kingdom of Singapura in 1947. In this universe, the British were expelled in 1823 and 1824, and the Malay rulers were in control.[13] Of interest to us is that the Malay and Sanskrit form of the name is used. This highlights the fact that the English (or, if you like, the colonial) form of the name has been retained, and this provides us with a contrast to Melaka/Malacca mentioned above and already points toward a pattern that will become apparent when other kinds of names are also considered.

Street Names

For this discussion of street names, I will only be concerned with official names, although it is possible that official names can be challenged by other names, as I have discussed elsewhere.[14] The pioneer of this work in the region, Raja-Singam, notes that these names are on people's minds all the time: "Postal employees have their eyes on them. Ambulance and fire-engine drivers take their direction from them as they rush by. Taxi-drivers and rickshawallas know them all."[15] The systematic naming of streets could be seen as a contribution of the colonial government. Raffles, the founder of modern Singapore, was also closely involved with town planning matters. In November 1822, he gave instructions to the planning committee that "each street should receive some appropriate name" and that it was "the duty of the police to see [that they were] regularly numbered."[16] This was successfully carried out although there were problems like duplication. Names were decided at municipal meetings and later at a special committee without consultation with the residents on the streets.

Street names in the colonial period generally employed English generics (Road, Street, Avenue, Crescent, and so on), although Malay generics (*Jalan*, "road, street;" *Lorong*, "lane") were also used, for example, Jalan Kembangan ("Expansion Road," named in 1932) and Lorong Limau ("Lime Lane," named in 1934). For context, these street names, including names with British associations, still exist. There is a general reluctance to change street names in Singapore; the position taken is that names should record the history of a place, including colonial history. A range of styles of naming were also employed. There were straightforward descriptive names like High Street, Circular Road, or Cross Street; there were associative names referring to objects and other entities in the vicinity, such as South Bridge Road, Orchard Road, or Church Street. The descriptive and associative naming styles are, in general, less ideological in nature.

There were also commemorative names with the inclusion of British personages or royalty in names, such as Mountbatten Road (named after the Supreme Allied Commander in the Second World War), Queen Street (named after Queen Victoria), and Queensway (named after Queen Elizabeth II). However, there are also streets named after local personages, such as Kheam Hock Road (named in 1923 after a municipal commissioner) and Meyappa Chettiar Road (named in 1932 after an Indian landowner). Commemorative names are more ideological in nature because they identify heroes or people to honor.[17] British place names (Bristol Road, Oxford Road) and place names from the region (Trengganu Street, Penang Road) were employed together. Streets were also named after cultural entities, flora and fauna, and abstract qualities, generally in English but possibly also in Malay (for instance, Butterfly Avenue and Bras Basah Road, where *bras basah* is Malay for "wet rice").

With Independence came instructions to "Malayanise" the street names, as Savage and Yeoh detail:

> In March 1967, ground rules laid down to guide the decisions of the Street-Naming Advisory Committee instructed the committee to steer clear of "old colonial nuances, British snob names, towns and royalty" and to give priority to local names. Preference was to be given to a fuller utilisation of Malay names to signal Singapore's allegiance to the Malay as opposed to the colonial world.[18]

It should be added that Malay is the "national language" (not just the official language) in Singapore, but currently, its function is largely ceremonial. Although there was talk of changing the names of existing streets, this did not happen, ostensibly for fear of causing confusion and inconvenience. However, new street names had to conform to the new rules. For instance, Chip Bee Gardens was constructed in this period and housed families in the British army. The street names were in Malay, with a color theme (i.e., Taman Warna, "Colour Garden;" Jalan Puteh Jerneh, "Clear White Road"). This is a clear move toward a decolonialized naming style.

This style of naming, however, did not always go down well. There were complaints about the many syllables in the names and that Malay names brought down the tone of housing estates. There was some attempt to circumvent this by employing "mathematical naming"—in other words, using numbers for naming streets. By May 1968, the Street-Naming Advisory Committee sent out a guideline to avoid using "Jalan" for "Road" and names of flora and fauna in Malay.[19] And so the experiment was not long-lasting, but these Malay-style names are littered around the landscape. Currently, the Street and Building Naming Board provides guidelines on street naming and approves applications for street names. The guideline now states that "Application for street names is only available in English. Only the UK English form of spelling should be used in street names. Non-English names such as Lorong and Jalan are to be avoided for new street names."[20] No justification is given for this choice, and perhaps the rationale is considered self-evident: English is accepted as (becoming) the most common language spoken in Singapore, and it unites the various ethnic groups here. The current range of street names in Singapore therefore show a range of naming styles, with names given in the colonial period largely remaining intact. New street names are in the English language, and there is no obvious break from the naming traditions in the (colonial) past.

Station Names

Like street names, urban transit station names also loom large in people's consciousness since they are necessary for navigating and getting around the city. By and large, street names are considered top-down names, meaning these names have been decided by government agencies with authority.[21] This is largely the case for station names for the urban transit, known as the Mass Rapid Transit (MRT) in Singapore. The first-generation names in the 1980s would be top-down names as they were decided by the MRT Corporation. The Land Transport Authority (LTA), which took over the function of the MRT Corporation later, allowed the names of stations for the later lines to be decided based on people's choices. They carried out a survey in 2004 for the names of the Circle line stations. This practice continued with subsequent new lines (the Downtown and the Thompson-East Coast lines). The public appeared to welcome this opportunity to have their voice heard. The process involved the LTA developing a working name for each station. There would be a feedback period of a month or so. The LTA would then provide a shortlist of three options that members of the public could vote for.

The other thing to note is that station names are a post-independence phenomenon. It is true that there were trams in colonial Singapore around the turn of the twentieth century, but these were all but forgotten come Independence. Theoretically, this meant that it was possible for the names to reflect post-Independence Singapore. Practically, however, for names to be acceptable to the public, these had to give a reasonable indication of location, which meant that a name based on the suburb or street could tend to be chosen. If a street name is used, there is the opportunity to reference one that is less colonial. Some of this seemed to have happened in the run-up to the opening of the stations. There was no question of avoiding names in English, though commemorative names will appear more colonial: "'We have a number of English names too,' Mr Fam [chairman of the MRT Corporation] said. 'Nothing against names such as Raffles, Lavender, Novena and Orchard. We have no hang-ups about our colonial past. But we thought we'd have a few more local names.'"[22]

This was the chairman's comment in 1986 in relation to the naming of the first-generation stations, which did not involve a consultation process. There was nonetheless a working name for each station, but before the stations were opened, the decision was made to replace the working names. In this case, the decision was to replace the name Victoria (commemorating the Queen) with Bugis (a people group from Sulawesi), and Maxwell (commemorating a British administrator) with Tanjong Pagar (Malay for the Cape of Fence). (Ironically, one of the stations in a newer line is named Maxwell.) But even in the original lines, there were stations called Raffles (after the British founder of Singapore) and Somerset (the English county). The strategy was to avoid sounding too colonial rather than to avoid sounding colonial at all. It is difficult to confidently say why the MRT Corporation made these last-minute changes. The public had not complained about British-themed names or English names. It is possible that Mr. Fam and other members of the Corporation were aware of name changes going on in neighboring Malaysia and other parts of the postcolonial world. This could be a manifestation of the *via media* or middle way of accepting some colonial names but not too many.

Lim and Perono Cacciafoco have undertaken studies that compare the names of the first-generation lines in contrast to the more recent MRT lines.[23] The names of the stations represent a mix of different languages, with English being a prominent language. For the first-generation North-South line, 11 (40.7%) out of the 27 stations had pure English names. (In other words, names that represent more than one language are excluded.) For the Downtown line (opening in stages between 2013 and 2017), 17 (50%) out of the 34 stations had pure English names. It would appear that the public, with the choice given to them in the consultation process, has continued to favor English names even more strongly than the MRT Corporation in their choice of names for the North-South line. This pattern of favoring English names is apparent when examining the names of the stations for the Thompson-East Coast line (currently opening in stages). Table 8.1 summarizes the working names, proposed names for polling, and the final names of selected stations where the proposed names included different language options.

We will notice that Maxwell, the name replaced in the North-South line, now appears in the new line. Lim and Perono Cacciafoco attribute this to the linguistic identity of Singaporeans as English speakers and to this choice as being pragmatic. Could there have been other reasons for the public's preference for English names? English is often thought to be useful for attracting international business and tourists. It seems unlikely that the public was thinking of people

Table 8.1 Selected working names, proposed names, and final names of stations on the Thompson-East Coast line (adapted from Lim and Perono Cacciafoco).[24] C = Chinese (any variety), E = English, M = Malay.

Working name	*Proposed names*	*Final name*
Springleaf (E)	Springleaf (E) Nee Soon Village (C+E) Thong Soon (C)	Springleaf (E)
Lentor (M)	Lentor (M) Lentor Green (M+E) Teachers' Estate (E)	Lentor (M)
Mayflower (E)	Mayflower (E) Kebun Baru (M) Ang Mo Kio West (C+E)	Mayflower (E)
Sin Ming (C)	Sin Ming (C) Bishan Park (C+E) Bright Hill (E)	Bright Hill (E)
Napier (E)	Napier Rd (E) Taman Serasi (M) Botanic Gardens South (E)	Napier (E)
Orchard Boulevard (E)	Orchard Boulevard (E) Tanglin (C) Grange Rd (E)	Orchard Boulevard (E)
Great World City (E)	Great World (E) Kim Seng (C) River Valley (E)	Great World (E)
Havelock (E)	Havelock (E) Bukit Ho Swee (M+C) Zion Rd (E)	Havelock (E)
Maxwell (E)	Maxwell (E) Ann Siang Hill (C+E) Neil Rd (E)	Maxwell (E)

from outside Singapore, rather than just considering Singaporeans. The public is likely to be less sensitive to the politics or optics of employing English names (whereas Mr Fam and the MRT Corporation, discussed above, might have been more sensitive to it).

Building Names

Residential buildings invoke a category of names that would be considered bottom-up names in Landry and Bourhis's perspective: these names are private names that have not been chosen by government agencies and could reflect commercial interests.[25] This is mainly a post-independence phenomenon, although it is true that the grand old houses in the past were named in the British style, for instance, Bendemeer House or Tyersall House, both of which were the source of the eponymous street names.[26] In this section, I refer to data that I gathered in an earlier article.[27] The names refer to the names of privately developed flats, known as condominiums or condos in Singapore. Most government-built flats are given block numbers and not names. These names not only represent the identity of these condos but are likely to be part of the marketing strategy of the condo developers. These names are chosen by developers and submitted to the Streets and Building Names Board for approval. There are probably many naming strategies employed, and saleability must be foremost in the developers' minds, so the name has to be attractive and aspirational to potential buyers.

The structure of condo names is similar to that of street names in that there is usually a specific element (which specifically identifies the object) and a generic element (which identifies the *kind* of object being named). Therefore, for the condo name Niven Loft, *Niven* is the specific element, and *Loft* is the generic element. In more recent names, the generic element might be abandoned altogether (as in the condo Vida), or might be replaced by other words (such as *The* and *Modern* in the condo The Lincoln Modern; and *Holland* in the condo Viz @ Holland, where *Holland* refers to the Holland Road location). For this reason, I use the labels "main element" and "other elements" in classifying names.

Like the classification of MRT station names above, I focus on the languages represented in the names of condos. I take the position that names can be said to belong to languages in that they arise out of particular language communities, and the orthographic and pronunciation rules are those from that language. For this reason, I analyze Niven Loft and Oxford Suites as having English main elements, although they are derived from a personal name (Lawrence Niven was superintendent of the Botanic Gardens in the 1970s) and a place name Oxford (a city in England, closely associated with the university of the same name). The condo names were retrieved from the Condo Directory website, describing itself as the "One-stop residential Singapore property guide."[28] A total of 2,599 names were obtained on January 31, 2008. The year when the temporary occupation permit (TOP) was issued was also available. (This is issued when a building is

Table 8.2 Percentages of the main and other elements of building names in English in different periods.

Period	*Main element*	*Other elements*
P0 (undated; presumed older)	66.3%	92.5%
P4 (before 1980)	87.2%	95.7%
P3 (1980s)	70.6%	90.1%
P2 (1990s)	71.2%	87.1%
P1 (2000 onward)	73.0%	77.0%

deemed habitable.) This made it possible to categorize properties based on their TOP years, and from 2000 onward, are labeled P1, those in the 1990s are labeled P2, those in the 1980s are labeled P3, and those before 190 are labeled P4. Those without TOP years are labeled P0, which are assumed to be older buildings.

About 37.2% of the condo names are based on street names. Given the dominance of English in street names, the importance of English for condo names is not unexpected. However, this still leaves 62.8% of the names for developers to come up with. In all the periods examined, English-based names formed the clear majority. This is true whether we are are concerned with the principal element or the other elements. The details are summarized in Table 8.2.

The percentages indicate the dominance of English in the condo names in Singapore. It is not surprising that the "other elements" bear a high percentage of names in English, as English is deemed the neutral language for the purposes of description. Even so, the percentages seem to have decreased in more recent periods.

For the main element, names from the other local languages of Singapore may have been employed (Malay in Duku Apartment [*duku*: Malay, "a topical fruit"], Chinese in Eng Hoon Mansions [*Eng Hoon*: Hokkien Chinese, name of a nineteenth-century Chinese merchant]). Among the reasons for a recent reduction in the English percentage for the main element are:

- the rise of newly coined names such as Viz, Mandale (probably an adjustment of Mandalay Road), Inspira (probably from *inspire* with a European ending added), Sensora (probably from *sense* with a European ending added), Montebleu (probably a French-style adjustment of Minbu, based on the street name Minbu Road), and Beccagayle (probably a portmanteau word: Becca + Gayle); and
- the use of continental European names, especially from French, Spanish, and Italian; examples include Papillon (French, "butterfly," based on the street name Jalan Rama Rama [Malay, "Butterfly Road"]), Alessandrea (the Italian form of Alexandra, based on the street name Alexandra Road), Costa del

Sol (the Spanish toponym, "sun coast"), and Tierra Vue (Spanish, "earth" + French, "view").

The reasons for the reduction in English percentage for the other elements are:

- the abandonment of the use of the generic element, as in Nuovo (a name with one element); and
- the use of continental European elements, as in Casa Uno (Spanish, *casa*, "house"), and Le Gambir (French, *le*, "the").

What is striking is the names appear to be moving away from local identity. Discussions of decoloniality generally focus on a return to autochthonous reference points. Not only is the English naming style dominant but names based on non-autochthonous languages like French, Spanish, and Italian have been adopted also. This might be understandable if we consider condo developers as selling a lifestyle like designers do. Many of the designer labels that we are familiar with are French (e.g., Louis Vuitton), Italian (e.g., Valentino), or Spanish (e.g., Manolo Blahnik). The continental European names are then seen as injecting a sense of class to the condo. A similar phenomenon has been noted in Hong Kong, with names like Chateau Vivienne, La Noblesse and La Hacienda. Jaworski and Yeung note:

> These languages [French, Spanish and Italian], being associated with some of the most powerful western-European nations and popular tourist destinations (for wealthy Hong Kongers), connote not only wealth but also . . ."high" culture, sophisticated, good taste and design, and a relaxed lifestyle.[29]

Personal Names

The final category of names I consider is personal names, a category well in the realm of Landry and Bourhis's "bottom-up" classification of names. In this section, I make use of data assembled in an earlier article.[30] Here, I focus on the names of the ethnic Chinese population, which makes up about three-quarters of the population of Singapore. The legal system in Singapore is based on Common Law (as opposed to Civil Law), and there are few restrictions on personal naming options. One point to note, however, is that name registration in Singapore has always been based on the Latin alphabet, as opposed to being based on Chinese characters, as is the case in China: registration is based on English in Singapore. For instance, in Singapore, you can register the name Lee Kuan Yew (with the usual options for spelling variation); in China, you can only register it in Chinese characters 李光耀. (This is not to say Chinese characters cannot be incorporated in the Singapore registration; the main form is the one in the

Latin alphabet.) In this sense, personal names have already been Anglicized for the requirement of an English-style orthographic form. Personal naming styles have the potential to spread or recede like all matters of style. De Klerk and de Klerk and Bosch discuss a trend toward the Africanization of personal names in southern Africa, with more preferring African, over English-given names. We see "increasing trends towards rejection of . . . 'English' names owing to increasing socio-political awareness."[31]

In Singapore, the government launched the Speak Mandarin Campaign in 1979 to unite the ethnic Chinese community who at that time spoke different Chinese varieties (mainly Hokkien, Teochew, and Cantonese) by getting them to speak Mandarin (the "standard" variety) instead. Part of the remit was also to persuade Chinese Singaporeans to use names based on the Mandarin pronunciation and to use the official Romanization for Mandarin known as *hanyu pinyin*. The prime minister at that time, Lee Kuan Yew, also talked about "mother tongue languages" including Mandarin as providing a "cultural ballast," for instance in a speech in 2004.[32] There was, therefore, a possibility of a movement toward the Mandarinization of names in Singapore, which could be seen as a Sinicization of the ethnic Chinese population, which would be the Singaporean counterpart of the Africanization movement in southern Africa mentioned above. Lee Kuan Yew's immediate family has practiced what they preached. Lee Kuan Yew himself abandoned his English-based given name Harry (given to him by his anglophile grandfather), before he returned to Singapore in 1950 after completing his law degree in Cambridge. His son, who was up to recently the prime minister, Lee Hsien Loong has a Chinese-based given name only. Lee Kuan Yew's grandchild Li Xiuqi (born in the early 1980s) also as a Chinese-based given name only, but this time in the official *hanyu pinyin* style, distinctive in its use of the letters <x> and <q>. The English-style spelling of the surname Lee has also been abandoned. This provides a clear example of how decoloniality of personal names could proceed.

Unlike in many other places, registered names are not publicly available. Alternative sources of name data had to be sought. Four sets of names from undergraduate graduation lists were employed: the 1960 list from Nanyang University (NU); the 1968 list from the University of Singapore (U of S); the 1983 list from the National University of Singapore (NUS); and the 2000 list from the National University of Singapore. It might be useful to note the NUS was formed in 1980 through a merger of NU and the U of S. The 2000 list excluded students in the faculties of Engineering and Computer Science due to a significant presence of students from China in these faculties. At this juncture, it will need to be mentioned that there are potential problems in using university name lists as a proxy for the whole Singaporean population. The choice of data was opportunistic. Nevertheless, I am confident that they can capture naming trends in Singapore as the data sets are comparable, and this is what I am attempting to present in this section. It might be helpful to note that these names represent those born between 21 and 25 years before the years of graduation. I noted the presence of

Table 8.3 Distribution of naming style options across different graduating cohorts. EGBN: English-based given name; CBGN: Chinese-based given name.

Graduating year	*University*	*With single EGBNs*	*With double EGBNs*	*Lacking CBGNs*	*With CBGNs in pinyin*
1960	Nanyang University	0.6	0.0	0.6	0.0
1968	University of Singapore	18.6	0.1	1.9	0.0
1983	National University of Singapore	16.0	0.5	1.1	0.0
2000	National University of Singapore	38.3	1.3	2.4	1.0

one or more English-based givens names (EBGNs), of Chinese-based givens names (CBGN). Table 8.3 summarizes the results in percentage terms.

There is the appearance of a kind of clawback with the 1983 NUS figures. This can be accounted for by the NU and U of S merging to form NUS. NU was a Chinese-medium university, and the U of S was an English-medium university, and the students would have more Chinese and English outlooks, respectively.

The overall pattern is for more Chinese Singaporeans to be given EGBNs at birth, with a growing possibility of multiple EBGNs being bestowed. There is also a slowly growing pattern of not bestowing CBGNs. The *pinyin* style for CBGNs is shown to be just appearing in the last row; it was only beginning to be known in the late 1970s in Singapore. The top EBGNs in the 2000 cohort are Sharon, Serene, Karen, Michelle, and Adeline for girls; and Kelvin, Terence, Adrian, Daniel, and Kenneth for boys. These names would appear in the standard list of given names in English-speaking countries. There are invented names (such as Lenita or Rayney) or the use of creative spelling (such as Georjina and Terrance), but these remain a small minority. There are also a small number of continental European names (Italian, Dutch, and German names: Mario, Nelleke, and Uschi, respectively). This is clear evidence of the Anglicization of personal names in Singapore, contrary to the pattern in Lee Kuan Yew's family. However, the use of Mandarin and *pinyin* is picking up, and the figures point toward a Mandarinization of names, but not convincingly because these were the early days of *pinyin*. It is worth noting that there have been discussions of the use of EBGNs in other ethnic Chinese communities outside of Singapore too, including Hong Kong and Taiwan.[33] Mainland Chinese overseas are also known to give themselves EBGNs (as discussed by Sercombe et al.).[34] It is unclear whether the ethnic Chinese have a cultural predisposition to be more accepting of EBGNs.

The Singaporean Perspective of Colonialism

We have been considering four categories of names in Singapore, both top-down and bottom-up names: street names, station names, building names, and personal names. We indeed see some evidence of an attempt to establish different norms to establish Singapore's new identity after Independence. These include the Malay-style street names in the late 1960s, the deliberate avoidance of using Victoria and Maxwell as station names, and the attempt within Lee Kuan Yew's family to adopt Mandarin Chinese-style names. What is striking is that these were not all-out attempts but that these attempts also did not last. When Singaporeans were allowed a choice, they went for naming styles that were not divergent from names established in the colonial period. The wholesale renaming process did not happen in Singapore. This might be attributable to inertia: better the devil you know than the devil you don't. There must, however, be more to this than that. I suggest three points that might further account for the naming phenomena that I have outlined before, suggesting that there has been a change in perspective concerning names.

A significant point is the consistent view of British colonialism from the government's perspective. Here is the pronouncement of Singapore's first prime minister Lee Kuan Yew a few years after Singapore's independence and 150 years after the arrival of Raffles:

> Decolonisation is usually supposed to mean the orderly dismantling of an empire. Conversely, viewed from the position of the subject peoples, it should mean a restoration to freedom and nationhood. But several hundred years of empire have created civilised communities where none previously existed. When Stamford Raffles came here 150 years ago, there was no organised human society in Singapore, unless a fishing village can be called a society. There are now over two million people with the second highest standard of living in Asia.[35]

British colonialism had brought with it society, civilization, and wealth. It had brought with it advantages, and there was, therefore, no interest in erasing or forgetting the colonial episode.

In 1984, Rajaratnam, the deputy prime minister at the time, talked about this again. He took the same position as Lee Kuan Yew in saying there was no real history before the arrival of Raffles, but he did go on to say that there were some negative aspects if imperial rule.

> Raffles founded Singapore. This is a fact. It is also a fact that the British ruled Singapore for 146 years. We may detest imperial rule, but as with Roman imperial rule in Britain, British imperial rule had both positive and negative aspects. It was both oppressive and liberationist. . . . We must take care not to throw the baby out with the bath water. But unfortunately in many former imperial territories the new rulers have thrown out the baby while retaining the bath water. . . . That is why in Singapore we have not tried to deny our imperial past by blindly eradicating both its constructive and destructive

> elements. We have not gone around renaming streets simply because they remind us of our imperialist past. This is to deny history.[36]

His comment at the end of the quotation is of special interest to us with an interest in names, and he explicitly explains the reason for not going on a large-scale name change process, so "Singapore" itself has been left intact as were street names bestowed in the colonial period. Names are also seen to present not only current identity but also historical identities.

In 2019, Singapore chose to mark the bicentennial of Raffles' landing with a series of events throughout the year. An interesting development that seemed absent in the 1969 and 1984 speeches is the acknowledgment that was actually something in Singapore before Raffles, and this was strikingly seen in the erection of four other statues alongside that of Raffles at the river front to celebrate other contributors to early Singapore history: Sang Nila Utama, Naraina Pillai, Munshi Abdullah, and Tan Tock Seng. There is possibly an eye cast at the troubled situation at the time in Hong Kong, and Singapore was contrasting its situation by highlighting its mixed heritage. British heritage is also mentioned by commentators such as Holden and by Tommy Koh, a Singaporean diplomat associated with the Bicentennial Conference, in his commentary.[37] He went on to describe the good and bad points of colonial rule.

> In my book, I described British rule as 60 per cent good and 40 per cent bad. The good things are well known. What were some of the bad things? The British were racists and treated us as second class (Eurasians) or third class citizens. The British promoted the smoking of opium in Singapore. There was neither democracy nor human rights in colonial Singapore. The British did not diversify the economy and were unable to generate enough jobs for our young people. Most egregious of all, the British failed to defend Singapore against the Japanese invasion. However, unlike other colonial masters in Southeast Asia, the British did leave us with a positive legacy. This includes the English language, the free port, free trade, open economy, good infrastructure, the rule of law, a good civil service and police force, town planning, public hygiene and modern medicine, a belief in science and modernity and an appreciation of nature and natural history.[38]

The view of British colonialism in Singapore thus appears more nuanced now. However, it is telling from the percentages given at the beginning of the quotation that it is still seen as an overall positive force. With this view of British colonialism, it is no wonder that there is no strong interest in dismantling the naming styles established in the colonial period. As noted above, it should also be remembered that the colonial government already had a more inclusive naming style that allowed for Malay-style names and the commemoration of local people in street names; the "local" was already inscribed into the landscape.

The Role of Englishization

Another factor for Singapore's continuing the trajectory set in the colonial period is that the process of language shift to English has made it completely natural to adopt practices associated with the English language. I use the label Englishization, rather than Anglicization, in the way employed by Kachru—how the English language has influenced languages and practices worldwide.[39] English vocabulary, grammatical structures, or discourse patterns might be borrowed from various languages. Kachru also refers to

> the power and influence of English in the domains of education, administration, literary creativity, and in international and intranational interactions as well, for it presents new perspectives and attitudes about English: English opening the doors to religious and cultural "enlightenment," English as a tool of colonial exploitation and political consolidation, and English symbolizing the "killer language" for various regional languages and cultures.[40]

In other words, Englishization can also be seen in how English takes control of domains, and more people prefer to take up and speak English. This Englishization of Singapore can be seen in the changing figures for the main home language, as seen in Table 8.4.

The figures for English have been steadily increasing since 1957 and are now Singapore's most common home language. The figures for Mandarin Chinese have also been growing, clearly replacing other Chinese languages, but peaked around 2010 before dropping. All of this could be seen to be precipitated by the medium-of-instruction changes in schools earlier. All Singapore schools became English-medium schools by 1987.[42] With English becoming the most popular language in Singapore, it will necessarily follow that the older attitudes toward English will change. These include notions of English being a colonial language or being associated with posh or highly educated people. The idea of English being a colonial language is fossilized in how it can be referred to in Hokkien as *ang moh oe*, "red hair speech" (as opposed to the more neutral *ing bun*, "English

Table 8.4 Speakers of the main languages in Singapore based on the language most frequently spoken at home in percentage terms.[41]

Language/year	*1957*	*1980*	*1990*	*2000*	*2010*	*2020*
English	1.8	11.6	18.8	23.0	32.3	48.3
Mandarin Chinese	0.1	10.2	23.7	35.0	35.6	29.9
Other Chinese	74.4	59.5	39.6	23.8	14.3	8.7
Malay	13.5	13.9	14.3	14.1	12.2	9.2
Tamil	5.2	3.1	2.9	3.2	3.3	2.5
Others	5.0	1.7	0.7	0.9	2.3	1.4

language"): *ang moh* refers to whites, and in particular the British (and therefore, *not* our language). English is now seen as a Singaporean language.

Ashcroft et al., in their book *The Empire Writes Back*, discuss postcolonial literatures written in the language associated with the empire, but with a difference, and this provides valuable insight to the changing attitude toward the imperial language. They describe a process of claiming ownership of the imperial language so that it is "ours" rather than "theirs."[43] This is achieved through the processes of abrogation and appropriation. Abrogation refers to how the imperial norms are rejected in favor of new norms suitable to the needs of the new speakers. This allows the language to be appropriated—accepted as belonging to the new speakers. Schneider's dynamic model of the evolution of English talks about the process similarly in his discussions about nativization and endonormative stabilization.[44] Singapore now celebrates its more local form of English, popularly known as Singlish, which exists alongside the more standard Singaporean variety. This means that English is perceived as Singaporean. English-given names also lose their foreignness. Singaporean English takes its place with other World Englishes. In fact, English surnames might also be appropriated as given names in Singapore, in the case of a baby named Beckham, after English footballer David Beckham.[45] It could be argued that Singapore has "tamed" the Englishness of the English language and even of English personal names, and these are accepted as Singaporean through the process of Englishization. Because English is now considered a Singaporean language, it is not felt to be unnatural to bestow new names in English and the English style.

True or Authentic Name vs. Postmodernism and Metroethnicity

"Authenticity" is a buzzword heard in various contexts, such as marketing, fashion, and art. It might be worthwhile considering whether names can be authentic.[46] An authentic name is said to represent the identity of the entity accurately. In religious and mythological contexts, there is the notion of the "true name." This can be seen in a few instances of renamings in the Bible, such as from Abram ("high father") to Abraham ("father of a multitude"), where the latter is said to be a better name and therefore a "true name," as it describes the patriarch's future. It should be noted that not all instances of renamings in the Bible have to do with authenticity. For example, Daniel is given the Babylonian name Belteshazzar when he is transported to Babylon (Daniel 4:8), and the name now instead appears to obscure his Jewish identity and is therefore "inauthentic." There are also traditional beliefs about how particular languages or styles represent particular ethnicities or nation-states. Therefore, it is deemed to be a certain correctness to the bestowing of Singaporean names to Singaporean people and entities. This runs into two problems. First, the notion of the Singaporean name itself is problematic: can we be sure what counts as a Singaporean name, given the mixed background of Singapore? Second, this belief can be dismissed as essentialist and unhelpful. These relationships are questioned in postmodernist thought. The British sociolinguist Maher proposes the principle of "cool" as the "main operating principle of cultural hybridity."[47] He suggests that ethnicity is not static but is malleable, hybrid and even contradictory and christens this "metroethnicity." Individuals are therefore not forced to only identify with their heritage culture when there is a vast smorgasbord available for choosing. A Chinese person, for instance, is not required to eat Chinese food all the time and might in fact sample or even prefer French, Italian, British, etc. food. Instead of being "inauthentic," choosing a French or Italian condo name might be deemed "cool." This more postmodernist conception of identity allows for a wider choice of name styles.

The principle of "cool" is a particular strand within a more general postmodernist approach to names and other phenomena. Talking of geographical places, the sociolinguists Pennycook and Otsuji talk about how "cities become layered, places of linguistic sedimentation, with buildings, walls, streets and signs carrying the traces of the past, and present stories of migration, movement, worship and successful and failed businesses."[48] Cultural categories become very mixed and change through time. In the context of postmodernism, how can one insist on any purist position on the use of heritage names? Moving away from a heritage name or employing non-heritage languages for condo names and personal names in Singapore might therefore be seen as part of the performance of "cool." The adoption of French or Italian condo names, apart from having a commercial motivation, might also be attributed to the "cool" mindset.

Conclusion

Singapore presents some unexpected characteristics in relation to postcolonial names. It has apparently chosen not to take the position of resisting colonial names and replacing them. The Singaporean view and experience of colonialism and embracing the English language must be related to this. The breakdown of the rigid modernist account of identity into a more fluid postmodernist one must also be pertinent to the names found in Singapore. We began our discussion by highlighting the aim of postcolonialism "to question, dismantle and subvert the binary hegemonic constructions that have supported and justified colonialism" with the expectation that naming practices would change to reflect postcolonial nationhood.[49] Our data on street names, station names, building names and personal names did not appear to reflect this. This raises the question: what, if anything, changed in Singapore's naming practices following independence? Considering that Singapore has a new constitution and related symbols of state (flag, anthem, coat of arms and so on). Yet, it has refused to turn its back on the English language because it could serve the nation internally: the baby was not to be thrown out with the bathwater. Singapore also did not oust the British army during its independence and was glad to have the army around for some years to address security issues. These decisions in relation to the English language and the presence of the British army could have been made on ideological grounds, too, but Singapore was focused on how they could serve the needs of the new state. This is often described as a pragmatic approach, which in the context of a government is defined as "a willingness to introduce new policies or modify existing ones as circumstances dictate, regardless of ideological principle."[50] It disregards orthodoxies and political correctness.

As a result of state policies, English grew in importance and, interestingly, through its prevalence, became accepted as a Singaporean language, eventually functioning as a strong component of Singaporean identity. This means that what the English language signifies has changed through time; the symbolic value of English has shifted from being an emblem of colonialism to emerge as part of Singapore's unique identity. As such, the significance of English names has changed. Singapore has appropriated the English language and changed its terms of reference in making it part of the Singaporean identity.

Notes

1 John Stuart Mill, *A System of Logic, Ratiocinative and Inductive: Being a Connected View of the Principles of Evidence, and Methods of Scientific Investigation* (London: J. W. Parker, 1874).

2 Saul Kripke, '"Identity and Necessity," in *Identity and Individuation*, ed. Milton K. Munitz (New York: New York University Press, 1971), pp. 135–164; Richard A Coates, "Properhood," *Language* 82.2 (2006), pp. 356–382, 364.

3 Mirko Casagranda, "Introduction: We Need New Postcolonial Names," in *Names and Naming in the Postcolonial English-Speaking World*, ed. Mirko Casagranda (Trento: Tangram, 2018), p. 7.

4 Peter K. W. Tan, "Limits to Styling: The Case of Adjustments in Commemorative Names," *Language & Communication* 57 (2017), pp. 29–36; Esterino Adami, " 'Good Names' in Indian English: Name-Giving Strategies and Symbolisms in Jhumpa Lahiri's Novels," in *Names and Naming in the Postcolonial English-Speaking World*, ed. Mirko Casagranda (Trento: Tangram, 2018), pp. 139−154.

5 "Challenges to Nationalism in Language Planning: Street Names in Malaysia," in *Researching Agency in Language Policy and Planning*, eds. Gregory Paul Glasgow and Jeremie Bouchard (London: Routledge, 2019), pp. 61–83; "Names as Linguistic Capital," in *The Commodification of Language: Conception Concerns and Empirical Manifestations*, eds. John E. Petrovic and Bedrettin Yazan (London: Routledge, 2021), pp. 164−180.

6 Mariana Isa and Maganjeet Kaur, *Kuala Lumpur Street Names* (Singapore: Marshall Cavendish, 2015).

7 Rodrigue Landry and Richard Y. Bourhis, "Linguistic Landscape and Ethnolinguistic Vitality: An Empirical Study," *Journal of Language and Social Psychology* 16.1 (2017), pp. 23–49.

8 Tan, "Names as Linguistic Capital."

9 Department of Statistics (2021), "Population Trends, 2021 (Singapore)," https://www.singstat.gov.sg/-/media/files/publications/population/population2021

10 C. M. Turnbull, *A History of Singapore, 1819–1988* (Singapore: Oxford University Press, 1989); Mary Somers Heidhues, *Southeast Asia: A Concise History* (London: Thames and Hudson, 2000).
11 Victor R. Savage and Brenda S. A. Yeoh, *Singapore Street Names: A Study of Toponymics* (Singapore: Marshall Cavendish, 2013), p. 349.
12 Savage and Yeoh, *Singapore Street Names*, p. 365.
13 Kevin Martens Wong, *Altered Straits* (Singapore: Epigram Books, 2017).
14 Tan, "Challenges to Nationalism."
15 S. D. Raja-Singam, *Malayan Street Names: What They Mean and Whom They Commemorate* (Ipoh: Mercantile Press, 1939), p. 12.
16 C. B. Buckley, *An Anecdotal History of Old Times in Singapore* (Singapore: Singapore University Press, 1984 [first published in 1902]), p. 84.
17 Tan, "Limits to Styling."
18 Savage and Yeoh, *Toponymics*, pp. 17–18.
19 MMACNRS (Minutes of *Meetings of the Advisory Committee on the Naming of Roads and Streets*) File 2/67, Dewan Bahasa Files, Ministry of Culture (Singapore: National Archives), 29 May 1968.
20 Street and Building Names Board, *Handbook for Guidelines on Naming of Streets* (2017), p. 11, https://www.ura.gov.sg/Corporate/Guidelines/Street-and-Building-Names
21 Landry and Bourhis, "Linguistic Landscape."
22 Rav Daliwal, "Three MRT Stations to Be Renamed," *The Straits Times*, November 28, 1986, p. 15.
23 Shaun Tyan Gin Lim and Francesco Perono Cacciafoco, "Then and Now: A Comparative Historical Toponomastics Analysis of Station Names in 2 of Singapore's Mass Rapid Transit (MRT) Lines," *Urban Science* 4.3 (2020), article 37; "Naming Public Transport and Historicising Experiences: Critical Typonymies and Everyday Multilingualism in Singapore's Mass Rapid Transit (MRT) System," *Urban Studies* (forthcoming); "From *Pasir Ris* to *Pioneer*: Singapore's Mass Rapid Transit (MRT) Station Names in Relation to Its Identity," *Onomastica* (forthcoming).
24 Lim and Perono Cacciafoco, "From *Pasir Ris* to *Pioneer*."
25 Landry and Bourhis, "Linguistic Landscape and Ethnolinguistic Vitality."
26 Savage and Yeoh, *Singapore Street Names*, pp. 37, 389.
27 Peter K. W. Tan, "Subversive Engineering: Building Names in Singapore," in *Engineering Earth: The Impacts of Megaengineering Projects*, ed. Stanley D. Brunn (Amsterdam: Springer, 2011), pp. 1997–2011.
28 http://condo.singaporeexpats.com.
29 Adam Jaworski and Simone Yeung, "Life in the Garden of Eden: The Naming and Imagery of Residential Hong Kong," in *Linguistic Landscape in the City*, eds. Elana Shohamy, Eliezer Ben-Rafael, and Monica Barni (London: Multilingual Matters, 2010), p. 165.
30 Peter K. W. Tan, "Evolving Naming Patterns: Anthroponymics within a Theory of the Dynamics of Non-Anglo Englishes," *World Englishes* 23.3 (2004), pp. 367–384.
31 Vivian de Klerk, "Changing Names in the 'New' South Africa: A Diachronic Survey," *Names* 50.3 (2002), pp. 163–173; Vivian de Klerk and Vivian Barbara Bosch, "Naming Practices in the Eastern Cape Province of South Africa," *Names* 44.3 (1996), pp. 167–188; and by the same authors, "Naming in Two Cultures: English and Xhosa Practices," *Nomina Africana* 9.1 (1995), pp. 68–85.
32 Lee Kuan Yew, "Eastern and Western Cultures and Modernisation," Speech delivered at the third session of the China Scientists Forum on Humanities, Beijing, April 21, 2004, https://www.nas.gov.sg/archivesonline/data/pdfdoc/2004042102.htm
33 David C. S. Li, "Borrowed Identity: Signalling Involvement with a Western Name," *Journal of Pragmatics* 28 (1997), pp. 489–513; I-Chung Ke and Chiu-Yen Huang,

"Parents' Perspectives on Adopting English Names in Taiwan," *Journal of Multilingual and Multicultural Development* 37.8 (2016), pp. 849–861; Ivona Baresova and Marcel Pikhart, "Going by an English Name: The Adoption and Use of English Names by Young Taiwanese Adults," *Social Sciences* 9.4 (2000), article 60.

34 Peter Sercombe, Tony Young, Ming Don, and Lin Lin, "Non-Heritage Names among Chinese Mainlanders," *Names* 62.2 (2014), pp. 65–75.

35 Lee Kuan Yew, Speech by the Prime Minister at the banquet given by the Singapore International Chamber of Commerce to mark the 150th anniversary of the founding of Singapore, February 6, 1969. Available online: https://www.nas.gov.sg/archivesonline/data/pdfdoc/lky19690206.pdf

36 S. Rajaratnam, "Adaptive Reuse: Integrating Traditional Areas into the Modern Urban Fabric," Speech delivered at Shangri-la Hotel on Saturday, April 28, 1984, https://www.nas.gov.sg/archivesonline/data/pdfdoc/sr19840428b.pdf

37 Philip Holden, "Colonialism with Benefits? Singaporean Peoplehood and Colonial Contradiction," *Journal of Southeast Asian Studies* 50.4 (2019), pp. 632–644; Tommy Koh, "My Birthday Wishes for Singapore," 2020, https://cil.nus.edu.sg/publication/my-birthday-wishes-for-singapore/

38 Holden, "Colonialism with Benefits?" pp. 632–644; Koh, "My Birthday Wishes for Singapore."

39 Braj B. Kachru, "Englishization and Contact Linguistics," *World Englishes* 13.2 (1994), pp. 135–154.

40 Kachru, "Englishization and Contact Linguistics," p. 36.

41 Lim and Cacciafoco, "Then and Now;" Department of Statistics, *Census of Population 2020: Statistical Release 1: Demographic Characteristics, Education, Language and Religion* (Singapore), https://www.singstat.gov.sg/publications/reference/cop2020/cop2020-sr1/census20_stat_release1

42 S. Gopinathan, "Language Planning in Education in Singapore: History, Transitions, Futures," *ASCD (Singapore) Review* 12.2 (2004), pp. 14–23.

43 Bill Ashcroft, Gareth Griffiths, and Helen Tiffin, *The Empire Writes Back: Theory and Practice in Post-Colonial Literatures* (London: Routledge, 2002).

44 Edgar Schneider, *Postcolonial English: Varieties around the World* (Cambridge: Cambridge University Press, 2007).

45 Charlene Chua, "Actress Pan Ling Ling Gets Beckham to Meet Her Son, His Namesake," *AsiaOne*, November 21, 2014, https://www.asiaone.com/entertainment/actress-pan-ling-ling-gets-beckham-meet-her-son-his-namesake

46 Peter K. W. Tan, "Naming as Styling: Inauthenticity in Building Names in Singapore," in *Language in Place: Stylistic Perspectives on Landscape, Place and Environment*, eds. Daniela Francesca Virdis, Elisabetta Zurru, and Ernestine Lahey (Amsterdam: John Benjamins, 2021), pp. 167–188.

47 John C. Maher, "Metroethnicity, Language, and the Principle of Cool," *International Journal of Sociology of Language* 175–176 (2005), pp. 83–102, 89.

48 Alastair Pennycook and Emi Otsuji, *Metrolingualism: Language in the City* (London: Routledge, 2015), p. 144.

49 Casagranda, "Introduction," p. 7.

50 David Seth Jones, "Governance and Meritocracy: A Study of Policy Implementation in Singapore," in *The Role of the Public Bureaucracy in Policy Implementation in Five ASEAN Countries*, ed. Jon S. T. Quah (Cambridge: Cambridge University Press, 2016), pp. 297–369, 316.

Index

Page numbers in **bold** indicate a table on the corresponding page.

Kolibri 360 GmbH
Ferdinand-Jühlke-Straße 7
99095 Erfurt, Deutschland
produktsicherheit@kolibri360.de